Wings Across America

The Celebration of a Miracle

by

Michael A. Rencavage

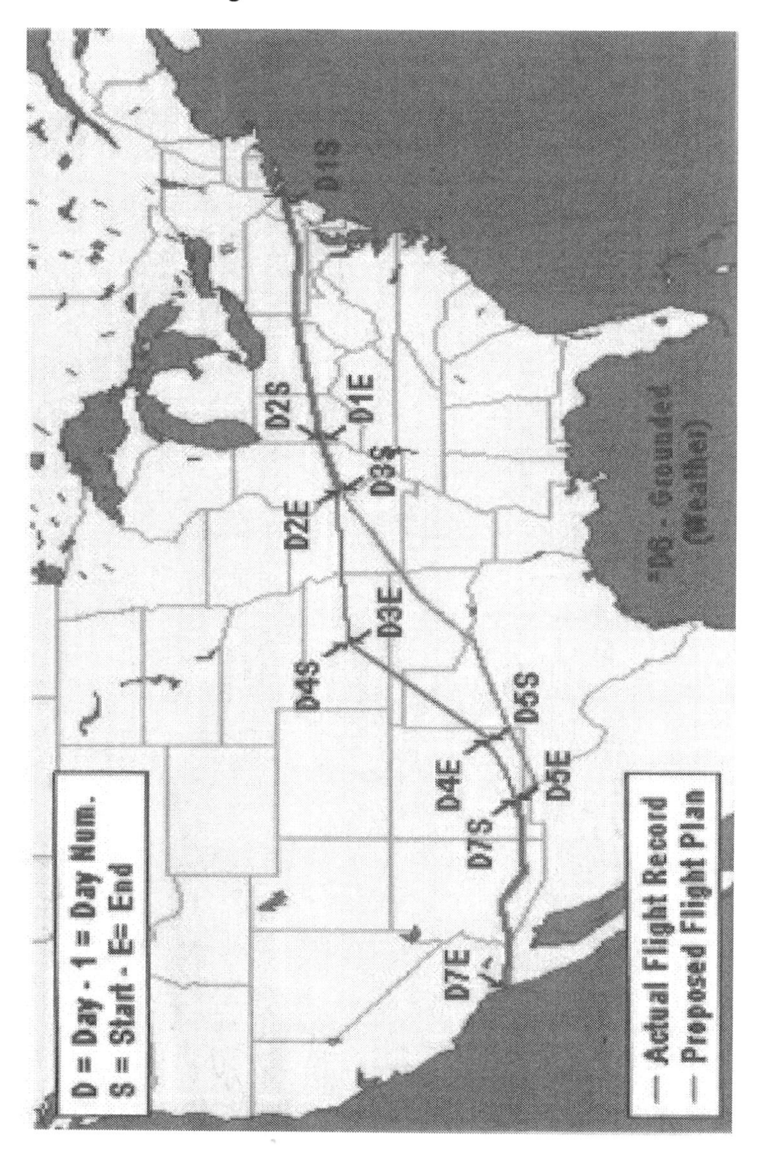
D = Day - 1 = Day Num.
S = Start - E= End
D1S
D1E
D2S
D2E
D3S
D3E
D4S
D4E
D5S
D5E
*D6 - Grounded (Weather)
D7S
D7E
Actual Flight Record
Proposed Flight Plan

This book is dedicated to our mother Charlotte Rencavage. A woman who could find the good in anyone and the silver lining to every cloud. Not a day goes by without a fond memory of her.

Prologue

Our predawn gathering had proven a bit too early for the local press. Only those dedicated members of the Wings Across America team and our ever supportive family and friends would bring life to the darkened ramp as we made our final preparations for departure.

The morning fog was still quite thick, a vestige of Hurricane Isabelle left to linger in the calm cool air of the last days of summer. As the sun rose slowly over the eastern hills, it illuminated the mist, and carried with it our hopes for improving skies and for departing on our journey, our grand adventure, so to speak.

With the passing of time, the weather would most certainly improve, but our waiting was tempered by our desire for progress. Wait too long and we would lose any chance of making Terre Haute by day's end, not long enough, and we would fly off into the uncertainty of single engine instrument flight in low overcast conditions, incurring a far greater risk than I was willing to take with Kimberly on board.

We chose to listen to the cautious voice of reason and delay yet a little longer, allowing us a few extra moments of joking and encouragement with those gathered to see us off. After about an hour, the airport's weather computer was showing improving conditions to the east. The visibility was coming up, and the ceilings were rising to the point that flight under visual conditions would soon be possible. At the airport we were seeing some positive signs as well, and the time had come for us to leave. I decided to depart on an instrument clearance and then continue the flight by visual reference once we were far enough east and flying in better weather.

We said our final round of good-byes. My eldest son Mikey just flashed that priceless smile of his as I hugged him in his wheelchair. He would miss me; I knew that, but he generally handles my departures well. Christopher is a little different. When I hugged him, I could feel how heart broken he was. I held him close to my chest and kissed his forehead. I promised him that when he turned sixteen, we would repeat this adventure together. It is a promise that I intend to keep and pray he holds me to. Spending that kind of time with him and sharing such a grand memory would be a priceless gift. Chris stared up at me with his deep hazel eyes, and I shared in his sadness for a moment; then I smiled, vowing to see him in San Diego next weekend. He hugged me once more and ran to his mother's side as I escorted Kimberly to the airplane.

Christopher helpingus preparefor theflight.

Mixture--Full Rich
Carburetor Heat---------------------------------Cold
Primer---Two Pumps and Locked.
Avionics Master Switch----------------------Off

Battery Master Switch-- ----------------------On
Beacon------------------- -----------------------On
Ignition Switch-----------------------------------Left, for Start
Propeller Area-----------------------------------Cleared with a yell that returns my mind to the bygone days of aviation's glorious past, and with the push of a small silver button, the early morning silence was shattered by the smooth constant rumble of

the little Lycoming engine.

We taxied slowly from the ramp, and I looked toward Kimberly for one final glance of approval. She, too, was ready. We stopped at the departure end of runway 22 of the Wilkes-Barre/Scranton International Airport and completed the final items on our checklist. Then, with my own voice still ringing in the headsets, we taxied into position for take-off. A gentle push on the left rudder swung us around, and I eased the power toward the firewall. The engine instruments rose into the green arch of the normal operating range, the airspeed indicator bobbled to life, and we were off and running. I jockeyed the rudder pedals to hold the centerline as the airspeed rose, 50mph, 60mph! The stick came alive in my hand, and with the slightest touch of backpressure, our little Piper Cherokee stepped softly into the misty morning sky. I raised the nose to climb, then rocked the wings to wave good-bye to our entourage still gathered on the ramp below. Moments later, we banked gently to the east and disappeared into the clouds. Our journey had begun!

PART 1

"Often times our most magnificent adventures are founded in moments of hardship."

Michael A. Rencavage

Chapter 1

I remember the moment quite clearly. It is the kind of memory that marks a point in time, like where you were on September Eleventh, or for the older generation, the day that JFK was shot. It is, of course, of a far more personal nature, but the level to which it has affected my life was of equal or greater magnitude.

It was a crisp clear October day in the fall of 2000. My sister Kimberly and her fiancé Duane had come home for a visit from York, Pennsylvania. They had recently moved to York to start their first jobs as Physician Assistants, their long awaited reward for years of sacrifice and study. The idea of Kimberly coming home for a weekend was not at all unusual, but rather a welcome treat, and a plan was made to gather at my mother's house on Saturday afternoon.

The homecoming started out just fine with the normal round of hugs and hellos. We were sitting at the kitchen table together, while our mother served coffee, when Kim said she had something to tell us. She started out strongly, but soon her eyes were filled with tears and her voice began to quiver. She looked downward and spoke softly words that each and every one of us fears, "I have cancer."

It was as if that blue autumn sky had suddenly filled with cold, dark, rain clouds. With just those few words, the whole world became gray and dismal. Our hearts sank and we listened as Kim and Duane explained the details of the deadly disease that she would have to face.

Kimberly, a woman who had been running on average five miles per day, had suddenly found herself having chest pains and shortness of breath when she exercised. The symptoms persisted and her endurance decreased to the point that she could hardly climb the stairs to her bedroom after a long day at work. She knew that something serious was wrong. She made an appointment to see her physician, expecting a diagnosis of heart disease or diabetes, as those ailments have plagued our mother since she herself was in her forties. Kimberly sat in the examination room in fear of just such a diagnosis. Her mind raced in anticipation of the doctor's findings, yet she never imagined what was to come next.

The cardiac work up came back normal, and they drew the appropriate vials of blood sending her on her way. When she received the results from her lab work, both Kimberly and her physician knew the diagnosis. Her white blood cell count had gone through the roof. Now a certain increase in white blood cells would be normal if one were battling an infection or something of that nature, but with results that were nearly twenty times the normal value, Leukemia was the obvious diagnosis.

The chest pains, shortness of breath and fatigue were all linked to the excessive number of white blood cells that her body was producing. With so many white blood cells, a much larger cell in proportion to the red blood cell, the balance had been disrupted.

She was becoming symptomatic because of the inadequate amount of oxygen carrying red blood cells in her system. Her body was literally starving for oxygen during periods of exertion. If left untreated, the excessive amount of large white blood cells could accumulate in her internal organs, a condition known as blast crisis, and eventually shut them down leading to certain death.

Kimberly was sent for a bone marrow biopsy that would confirm the diagnosis by identifying a chromosomal anomaly known as the Philadelphia Chromosome. It did, in fact, rear its ugly head, and the confirmation was made that she was caught in a battle for her life against Chronic Myelogenis Leukemia, CML. The bad news was, of course, the diagnosis; the good news was that she had been plagued by the disease for only three months prior to being diagnosed. This would be a very strong positive in Kim's defense and a blessing for which we will always be grateful.

At first I was in disbelief. Of all the people I have ever known, Kimberly was certainly among the most fit. She was always cautious about her diet and exercised religiously. I realize now, however, that such factors make very little difference in the arena of blood related cancers. Nonetheless, I was momentarily stunned, but the disbelief faded quickly. I started to think like most men would. My first reaction was to address the problem at hand and start looking at the possible solutions. Instead of breaking down into tears, I felt we should start whatever treatment was deemed best as soon as possible. Don't waste any time; don't let the cancer get another moment's foothold. I didn't realize, of course, that my inner fears and emotions were being suppressed by my more practical side. That realization wouldn't come until later, but it would surely come.

The battle lines had been drawn, and it was time for Kimberly to choose her weapon. Initially she was given a prescription for a medication that would slow the progress of the disease and afford her some time to decide which form of treatment she would use to defend herself. At the time of her diagnosis, she was presented with three primary methods of treatment and informed by the experts as to the pros and cons of each.

The first method was a combination of Interferon and chemotherapy. Although this method has seen a decent amount of success, it was very hard on the patient's body and would make for a very difficult battle. She would only consider it if all else had failed. (Since the time of Kim's illness, there have been significant advances with Interferon which have greatly reduced the side effects and, as a result, the discomfort of this type of treatment.)

The second method was a brand new drug called STI571, now known as Glevac. At the onset of Kim's illness, this drug was in the final stages of FDA approval. It was a greatly simplified process with very few side effects. She would just take the medication daily and, if all went well, arrest the leukemia with the possibility of curing it. This seemed to be the answer at first. Through further investigation, however, it was found that in order for Kimberly to qualify for the drug, she would have to volunteer to become part of a test group. This group would be divided into half, with one-half receiving STI571, and the second group receiving Interferon. There were no guarantees as to which group she would end up in. It was also found that in most cases, STI 571(Glevac) did an excellent job of holding the leukemia and its symptoms at bay, but it was not often curative. Stop taking the medicine, and you would very likely end up where you started. There was no real guarantee that over extended periods of time the disease would not continue to progress, regardless of the medication. It was at the time a brand new medicine. These were concerns worth pondering. For Kimberly, Glevac did not prove to be the best possible alternative. It is, however, a grand breakthrough in the battle against leukemia which has saved many lives and will continue to do so in the future.

The third option, the one she would eventually choose, was that of a bone marrow transplant. It showed the potential for a full recovery and, overtime, the return to a normal lifestyle. The transplant process itself is rather long and best taken one step at a time. It is indeed taxing on the body and will surely test the resolve of even the toughest of individuals. Yet each step is a step toward potential recovery. It would not merely arrest the symptoms; instead, it would provide a cure. The catch to the bone marrow transplant is that its greatest chance for success occurs only if you have a bone marrow donor that is a genetic match. Oftentimes such a match is very difficult to find, even among the family of the patient.

For Kimberly, this would turn out to be yet another blessing worth counting, as both my brother Joe and I qualified. We had not one, but two matches, both willing and able to help, both with a greater than average fear of all things medical (at least when it came to being the patient), but both willing to do anything that would give our sister the second chance at life that she so deserved.

After some careful consideration Kim and Duane had decided in favor of the bone marrow transplant. They knew they would be facing a rather difficult time and began the preparations immediately. The main concerns revolved around Kimberly's immune system; during the process it would be completely defeated. This would require some planning on their part to help prevent future unnecessary battles with infection. Many of the concerns would never have crossed my mind had it not been for Kimberly's doctors bringing them to her attention. She would no longer be able to wear contact lenses, so she had to be fitted for eye glasses. Dental work would present a great risk of infection, so all of her dental needs were looked into carefully and addressed prior to her starting the transplant process. Dozens of little details needed to be attended to, and that is how they would spend the majority of their time over the next few months.

There was at least one exception to all of this, a little thing they felt needed to be addressed sooner rather than later. Kimberly and Duane had gotten engaged the previous spring, with plans of marriage sometime in June of 2001. They decided instead to move things up to November. Within months of her diagnosis, in the midst of the preparations for her transplant, I found myself escorting my sister down the isle to the man that she loved more than anyone in the world.

Duane truly loves my sister as well. He professed his love on the day they got engaged and has never once looked back. When a lesser man would have run, Duane chose to stand strongly by her side. Faced with adversity, he charged forward, moving up the wedding date and affording Kim the opportunity to enjoy fully the most memorable day of her life. The decision to show such love and commitment is something that he will never regret, for in that decision he chose a life that stretches far beyond the bonds of her illness. As a result, they have earned the rewards of a relationship founded in love and tempered by hardship, a mighty sword that shall defend them against the perils of everyday life. Their wedding was the public proclamation of their union, and I was truly honored to be a part of it.

It was a relatively small gathering of family friends. The ceremony and reception were held in a beautiful French Manor overlooking the rolling hills of the Pennsylvania countryside. Their wedding brought a glimmer of hope to the dark and fear-filled winter that lay ahead. It was a joyous gathering of the people who loved Kimberly so dearly at a time when we needed the support for which a large family is often known.

I remember sitting alone at the table watching as Duane and Kimberly danced. Their eyes gazed softly at each other; they smiled, caught up in the magic of the moment. The candle in front of me flickered, and my thoughts drifted to the little girl inside the woman. My mind flashed through scenes from our childhood, scenes of Kimberly and me walking to school together and playing in my grandmothers backyard. I remembered her smile at Christmas and the tears we shared when our father died. But, most of all, I remembered her compassion. Even as a little child Kimberly's compassion defined her. She was always the one concerned with the feelings of the other kids, always trying to make things just a little better for the rest of us. I remember one instance when I was eight years old. I was dying to take karate lessons with my best friend Bob Charette. My parents couldn't afford the uniform, much less the monthly dues. I would tag along with Bob and his dad and watch the classes, going home and practicing on my own. When Kimberly saw what was going on, she saved her money from birthday and Christmas gifts for six months just to buy me a gi (uniform) and keep my dream alive. That was quite a feat for a nine year old little girl. I still have that gi. My son Christopher uses it when we practice karate together.

It's funny how they say the eyes are the windows

to the soul. In Kimberly's eyes that evening the compassion still glowed. It glowed in the soul of a beautiful and intelligent woman, a woman of passion and perseverance, a peaceful warrior undaunted by the battle she was facing. I sat there that evening reflecting in the glow of that flickering candle, and then I paused and offered God a silent prayer of thanks and a petition for his continued blessings in our family's time of need.

Chapter2

Christmas and New Years Day came and went that year. The holidays passed with maybe just a little more love and appreciation for the time we had together. Throughout February I had made several trips to The Johns Hopkins Medical Center in Baltimore. They would test and retest me, verifying that I was, in fact, the most appropriate donor and in sufficiently good health for the procedure. I came through it all with a clean bill of health, no transmittable diseases, a good EKG and a stellar chest X-ray. That was a relief for me, because I wanted more than anything in the world to be able to help my sister, and also because I tend to worry about medically-related issues. I think that my propensity toward worrying about these things has its roots in my profession. I am a professional pilot by trade. Aviation is a hobby turned profession for me somewhere around my 33rd birthday. It is a wonderful thing to be able to make my living at something that I love to do. The only problem with my occupation is that any number of medical maladies that might be considered minor to a person in a different line of work could end my career and leave me without an income in the flash of a doctor's sympathetic smile. So, as you can see, it would be somewhat normal for me to be leery of allowing a group of medical professionals to poke and prod and test my body to such a degree.

It may sound a little ridiculous, but I remember keeping things in check by reflecting on all of the testing that the early NASA astronauts went through to qualify for the space program. What I was going through was nothing in comparison to the level of testing they had undergone, but those thoughts helped me to keep a positive outlook as I trudged through the halls to EKG, X-ray and the laboratories for the various tests that would eventually qualify my fitness.

We were still shaking off the cold as we entered Kimberly's hospital room on that brisk second day of March, 2001, the official day of the transplant. I couldn't get over the thought that she seemed too calm and relaxed. I thought she would be more worried, more anxious; after all, her life hung in the balance of a procedure that was beyond her control. But there she sat, as calm as could be, laughing and joking with her husband when we broke into the conversation to say hello. That is, I believe, the definition of grace, to hold your own and maintain your composure when faced with a situation that would leave others trembling. Kim sat up in her bed that morning, a vision of lucid grace, ready to face whatever might come.

The battle had already begun. Kimberly had

been given a treatment of chemotherapy that destroyed her bone marrow and its ability to produce the cells that were damaged by leukemia. The vast majority of that marrow was destroyed; the rest would be defeated over time.

The life giving blood of our bodies is in itself a miracle. When Kim was first diagnosed and as the transplant process continued, our entire family gained a layman's understanding of how the blood in our bodies is created and how the cells within that blood maintain the critical balance required to support life. It is a truly amazing process that I only understand on a cursory level, but it reminds me that the human body itself is one of the greatest miracles of all. I am not capable of explaining the things that go on at the cellular level during a bone marrow transplant, but, rather, I wish to give an understanding of the trials my sister faced in a long battle with a deadly disease.

Along with destroying her bone marrow, or more accurately as a result of the destruction of her marrow, the chemotherapy rendered Kimberly's immune system nonfunctional. Our visits became a masked, robed, scrubbed, sterilized gathering of family and friends.

After a short period of time, it was up to me to do my part. A nurse escorted me to the O.R. There we met with the anesthesiologist to discuss the various methods of numbing the physical pain of being a donor. I opted for the epidural, a simple and effective method that allowed me to avoid being put under unnecessarily. Ever since I had witnessed a few surgeries first hand, while doing a study of cardiac monitoring equipment in the operating room, I have always avoided being put under unless it was absolutely necessary. God bless those wonderful medical professionals for keeping us safe when we truly need them, but why take the chance of a possible complication if it isn't truly compulsory. Besides, my recovery room time would be less with my chosen method.

I then spent the next forty minutes or so undergoing a procedure that put two sets of guides or channels about 1/16 of an inch in diameter into the upper portion of my buttocks. They then inserted some rather large needles through the channels and into the back of my pelvic bone. The process was repeated literally hundreds of times as there is currently no efficient way of removing the marrow from one's bones. Now there's an invention I should be working on. The actual procedure was painless. The guides eliminated the need to repetitively puncture the skin and there are no nerves in the bone itself. The epidural insured that all that I felt was pressure. Afterwards I was very stiff and put on painkillers. The pain and stiffness were generated by the piercing of the muscle tissue that was nicked and irritated by the needles. I've been asked by several friends what the pain was like, and, truth be told, it was no worse than a sport's injury or something of that nature.

There is another item on my list of things to be thankful for. The amount of marrow required varies for each person depending on the ratio of usable marrow to what was extracted and the overall fitness of the person receiving the donation. I was in luck on both counts, thank God. My ratios were very good, and Kimberly was as fit and trim as she could possibly be. I'll have to remember to thank her for that as she greatly reduced the amount of time I had to spend on the operating room table.

I sat in the recovery room unable to feel anything from the waist down. While I waited patiently for my legs to return to working condition, the nurses lightened the situation with jokes and enjoyable conversation.

I continued my little visit with the recovery room staff while I waited for my body functions to return to normal and all of the appropriate boxes to be checked on the little list the nurse was attending to. Once it was completed I would be allowed to leave, but not before. We made a game of it, a race in fact, to beat my marrow to Kimberly's bedside. The marrow was being processed in the lab as I willed my body to return to normal as quickly as possible. In the end the marrow beat me to her room, but only by a few minutes. I still contest its victory for I was unfairly delayed, reduced to traveling the halls of the hospital by an ineffective means of transportation, the donor shuffle. The donor shuffle is a trademark mode of getting from one place to another just after completing the process of being a bone marrow donor. The nurses are quite familiar with the short shuffling strides that reminded me of the way Tim Conway would walk when portraying an old man on the Carol Burnett show back in the seventies. It took a lot of time to make a little progress, and it was pretty comical to watch.

When I got back to Kimberly's room, she was sitting up in the bed watching the marrow drip intravenously into her system. She was speaking to Duane, rather calmly discussing the next steps in the process. We visited for the entire first day, never really leaving her room, save the occasional run to the cafeteria. We stayed with her for the next two days as well while various family members and friends visited with warm wishes and prayers of support. A group of Kim's friends had gotten together and thrown a fund raising party to help defer the costs that Kim and Duane would personally incur and to offset the loss of her income. They raised several thousand dollars and had a great time in the process. We viewed the video with Kimberly, and she was nearly brought to tears by the outpouring of love and support that flowed from friends ranging from the present day to those from over twenty years prior. Kimberly was doing fine, and she persisted with what appeared to be a tremendous amount of inner strength.

At the end of the third day, we left Kim and Duane and departed for home. That evening is when the painkillers started to give me trouble. I am not very comfortable with feeling loopy, and I had no symptoms prior to leaving the hospital, but when we finally stopped to get something to eat, I was feeling very out of sorts. I was shivering and shaking with my brain lost in a fog of disorientation. I am not sure if it was my discomfort with the lack of control or another side effect of the medicine, but I felt as though my breathing was labored as well. I was in the middle of a reaction to the medicine designed to ease physical pain. Instead it was causing a great deal more problems than it could possibly be worth. At that moment I made a pact with myself that shall remain my credo for as long as I can possibly stand it. I shall choose pain over medication of any kind. I thought something was seriously wrong with me. While I lay there concentrating on my breathing and trying to relax, my mind went through a list of possible problems that ranged from the medicinal reaction all the way to an untimely heart attack.

The Emergency Room of an inner city hospital

is filled with all kinds of people. There sat my mother with the junkies and crackpots while I was escorted to the area where I would be tested and finally released. It was undoubtedly my lack of food and drink combined with some very strong painkillers that had caused the problem. I also feel that an underlying sense of concern for my sister's well being was a contributing factor that aggravated the situation. We spent that night at a nearby Holiday Inn and visited with Kim once more the following day before finally heading home.

I learned a very important lesson that day: to adhere to my personal policy of avoiding the use of any medication whenever possible. The physical pain was insignificant compared to the reaction I had to the medicine or the turmoil it caused for those around me. I can't imagine deriving pleasure from something that alters one's state of consciousness the way that drugs do. My heart goes out to those who are addicted to drugs, prescription or otherwise. One fleeting occurrence on a prescription pain killer was far more than I ever needed in my own life.

Chapter 3

Kimberly's immune system had been defeated by the chemotherapy and she spent the entire month of March in her hospital room with limited time roaming the halls and walking on the treadmill in the IPOP (Inpatient, Outpatient) wing. Kim would walk the halls dragging her I.V. pole to the room with the treadmill in it, then walk for thirty minutes to an hour at a time. It was a demonstration of pure perseverance. She was determined to do everything within her power to get better as quickly as possible.

Over the course of that month, the side effects of the chemo would begin to present themselves. Kimberly dealt with daily battles of nausea; she fought bravely with feelings of exhaustion and weakness. She eventually lost all of her hair. The hair loss was a psychological blow to her. If she didn't look into the mirror and think about her illness before, the loss of her hair would certainly take care of that oversight. Somehow, she still remained upbeat and positive during our visits. I thought about what she was going through and couldn't understand how she did it. I myself had recovered rather quickly from my minor role in the process only to find myself plagued with worry. It was the strangest thing, I somehow felt responsible for the results of the transplant. I couldn't stop worrying whether or not my bone marrow would do the job and save my sister. My concerns were haunting me. I was finding myself waking up in the middle of the night with gastro-intestinal pains that plagued me so much that I was getting apprehensive about lying down to go to sleep. It was one of the most stressful times of my life.

We would visit Kimberly as much as possible. In that first month we watched as my healthy, glowing, exuberant sister was transformed into a pale and feeble-looking version of her former self. The chemotherapy treatment had lingering effects, and her system was basically defenseless against the common sources of infection that a healthy immune system fights on a daily basis. It wasn't long, however, until she began to turn the corner, and the more that I would see her progressing, the better I began to feel.

In her second month post transplant Kim had to spend her time in The Hope Lodge, a sort of step down facility within minutes of the hospital. Living there kept her nearby just in case she were to encounter any further complications. It didn't take long for her to see the value of that precaution. During her second day at The Hope Lodge, her liver had started to swell, forcing an end to the limited liberation she was just beginning to enjoy. She returned to the hospital with additional worries that the chemotherapy may have permanently damaged her liver. By the grace of God and the helping hands of the Hopkins' medical staff, in just one week, her liver problem completely resolved itself.

Kim continued to progress, slowly at first, returning to The Hope Lodge with the 24 hour a day companionship of our youngest sister Cathy. Cathy really came through when Kimberly needed her the most. They were always close, but when Kim needed continual assistance, and Duane had run out of time off at work, Cathy took a leave of absence from teaching and stood by Kimberly's side. She spent her time carefully watching Kimberly for signs of rejection. Perhaps even more importantly, though, she provided Kim with a glimpse of normalcy as they routinely set out on little day trips and enjoyed a special bonding of their spirits that is still evident in the special relationship they share.

Shortly after that, Kimberly returned home with constant monitoring visits to Cancer Care Associates of York. She went about her daily activities adhering to all of the precautions necessary for her compromised immune system. In general, aside from the need for a mask in public and the wig she had to wear, she was slowly getting her life back. In September of 2001 Kim had a bone marrow biopsy. It provided the first significant indication that she was, in fact, being cured. Six months after the transplant, my bone marrow was taking over, and there was no sign of the dreaded Philadelphia Chromosome.

Kim spent the next year with overall good results; however, one problem seemed to persist for an extensive amount of time. Kimberly was constantly in need of blood transfusions. A procedure that would normally not be required beyond three to six months post transplant continued for more than a year. The transfusions were beginning to cause complications by introducing an excessive amount of iron into her system. She was put on medication and eventually had a device implanted to help regulate the iron, preventing it from causing additional
damage. I think we all worried about this to some degree, and eventually the doctors sent Kim back to Hopkins for another evaluation in October of 2002.

It was then that they discovered the Philadelphia Chromosome again, signaling a resurgence of the cancer. It was a tough blow to Kimberly, but she managed to stand strong as they presented her with her options. She was told of a Procedure known as a Donor Lymphocyte Infusion. (D.L.I.) It would act as a kind of booster shot to the transplant and hopefully evoke a controlled graph versus host (rejection) reaction. The D.L.I. would allow my marrow to take over completely and defeat the few remaining cancer cells from Kim's original marrow.

The Tuesday before Thanksgiving 2002 we went to Hopkins together. I was hooked to a machine that took the blood from my one arm and filtered out just the required "T" cells. It then returned the unused portion of my blood via my other arm. The whole procedure took just under an hour and then we waited as they prepped Kim and transfused those "T" cells via an intravenous drip.

Leaving the hospital that day, I felt overwhelmed with fear for Kimberly's future. We seemed to be running out of options. When we stepped through the doors and headed for the street, we were suddenly struck with the vision of a huge black sign with bold white letters. Its simple one word message was just what I needed. **"BELIEVE"** struck a chord that rang true for me like a message from the Almighty himself. It eased my mind, and the thought carried on throughout my day. When we returned home, a magazine we subscribe to had the title of its feature article plastered across the cover. "Believe In Modern Day Miracles." Later that day I opened a letter from a close friend that contained a magazine article; it carried the very same sentiment. I am not really sure what was going on in Baltimore that prompted those **"BELIEVE"** signs. We saw them in several places throughout the city. I will admit, though, that it bolstered my faith. When I told Kim of the varied presentations of the same message that I had received throughout my day, she paused for a second, then she shared with me the source of the great inner strength she had demonstrated throughout the ordeal of her illness. She told me that she could hardly sleep her first night in the hospital. She sat up in her hospital bed paralyzed by fear. The second day brought much of the same, but on that second night her life would be changed, and she would find the inner peace she needed to move forward without fear or anxiety.

That night, after Duane had left, Kimberly sat alone. She was scared to death with a million worries running through her mind. She started to dose off and awoke shortly afterward as the entire room grew luminescent, glowing in ice blue light from one end to the other. When she looked to her left she saw him. A strong and powerful presence, with one high arched wing folding gently across his chest and the other draped softly behind him. She had seen him before, his strong square jaw and broad shoulders. Once before, in fact, just as she was struggling through the darkest days of her divorce from her first husband, when she felt she could hardly go on. That's when he had first appeared to her. The second time she knew the trials would be far greater, but he left her that evening with the peace and inner strength she would need.

Her guardian angel was the reason she had shown such grace under pressure and such power to persevere.

It was a month or two before we got word that the reaction the doctors had hoped for had taken place. From that time forward, Kimberly's progress has been wonderful. She stopped needing transfusions and the tests, at the time of this writing, show no signs of the Philadelphia Chromosome. The marrow in her body is 100 percent from me, and she has returned to a normal life. With the help of God the process will continue to be the success the doctors expect it to be. We are very grateful to the dedicated medical professionals that have helped to facilitate the miracle of Kimberly's recovery, yet we remain cognoscente that it was only by the grace of God that she survived. We all prayed for a miracle, and I think it only fitting that we never forget the miracle we have been granted.

PART 2

"The important thing is to strive towards a goal which is not immediately visible. That goal is not the concern of the mind, but of the spirit."

Antoine de Saint Exupery
Flight to Arras

Chapter 4

Some ideas come to us slowly, over time, the gradual collection of pieces from a puzzle that fall gracefully into their appropriate places. The really good ones, however, tend to strike from nowhere, inspirations found in the first conscious thoughts of a new day or sparked by the words of another. They ignite the soul with flames that refuse to be quenched until given the careful consideration they so deserve.

It was the middle of May, 2003. I had been talking with Kimberly over the past several months about joining together to do something to raise money for The Leukemia and Lymphoma Society, the premiere agency in the battle against blood-related cancers. We had considered the possibility of running a marathon, but she was just getting back to normal, and that would be a daunting challenge for even the healthiest of individuals. We also felt that it would be difficult to raise much more than a few hundred dollars by participating in such an event. We had set our sights on raising a little more than that, if at all possible.

I had just finished re-reading an excellent book by Rinker Buck entitled "Flight of Passage." It is the tale of two young brothers and their adventurous flight across America in a Piper Cub set in 1966.

It was that book that inspired me. Just as I closed the cover it hit me. We could fly across the country together in a small single engine airplane, the bone marrow donor and the transplant recipient. It would be a great angle for the press to utilize and allow our project the chance to make the significant contribution we desired. By the end of the hour, Kimberly was onboard with the idea and looking forward to our grand adventure together. We felt that by combining our desire to help The Leukemia and Lymphoma Society with a cross country odyssey that would take us from New York all the way to San Diego, we could complete our goals. It would be a celebration of life and the miracle of Kimberly's recovery. It was an inspired thought.

As with all thoughts, however, we needed a plan, for without a plan even the greatest of ideas will meet with its demise, lost forever in the realm of some days and could have dones. So there I was with my supposed big idea and absolutely no means of bringing it to life.

I have logged thousands of hours as a professional airline pilot, being paid to assure the safe transport of my passengers from point A to point B, but that was, of course, done in someone else's aircraft. This has been the case for all of the flying that I have done over the past seven years. I had sold my airplane in an effort to raise the money I would need to complete the advanced flight ratings required for my career. I closed the door on my life as a private aircraft owner in an effort to open the window on the field of professional aviation.

As you can see, I had this grand idea of crossing roughly 2400 miles of the greatest country on earth while fighting leukemia with my sister, but I had no aircraft in which to do it. I was also brand new to the idea of fundraising and didn't possess as much as a clue where to begin. All I knew was that I really would like to do this and so, I sat down that evening and prayed. I prayed for guidance and asked God that if it was his will for us to proceed that he guide us along the way.

I got exactly what I had prayed for. Within three days, I had an airplane, not just any airplane mind you, but my old airplane. Its new owner, my good friend Dr. David Speranza, had heard through Kimberly of our plan and had called to offer the use of his aircraft. Another friend, Doug Fink, had put his company Unetsco and his director of website development, Allan Glass, at my disposal free of charge. Later in the project, Doug would also put his computer and pilot skills to work for me by helping me do all of the preliminary flight planning. In just a few short hours he and I had accomplished a task that would have taken days had we done it the old-fashioned way. He was a tremendous help with the entire project including assisting with the fund raising efforts. All I had to do was ask, and Doug would do everything in his power to help. I wish he could have come along for the ride as he and I have shared in many adventures aloft over the years. We even had a name for our project, "Wings Across America," and the expert guidance of Ms. Beth Mohan of The Leukemia and Lymphoma Society to coordinate our fund raising efforts.

The old saying, " be careful what you wish for," had proven itself all too true once again. It was the middle of May when I conceived the idea of our fund raiser, and by June 1st we had the website up and running, ready to tell our story and take donations via credit card. John Millington of Tech Aviation had also joined the team and dug right in, seeking corporate sponsorship from his aviation contacts. During the flight I utilized his years of experience as a flight dispatcher. He helped in analyzing weather data and flight routes to raise the safety level of our journey westward.

Since the aircraft was based over 100 miles away, Dave was put in charge of aircraft preparation. He mailed me all of the literature he could on the new avionics and performance specifications to help me become reacquainted with an airplane that was more of an old and trusted friend. Over the months leading up to our departure I flew with Dave on a few different occasions. He patiently walked me through the subtle differences in the airplane over the years. His tone was that of a doting father as he explained: "Start her on the left mag only, with the alternator off," and "She runs better if you lean the mixture about one knuckle before starting to taxi." Although N7203J was an old and trusted friend to me, she was more like a member of the family to Dave. He continually applies the tender loving care necessary to keep his then thirty-five year old airplane looking and running like new.

I remember the first time I saw the little Piper Cherokee. She was sitting in the grass at the Tunkhannock Airport, a little strip in Northeastern Pennsylvania that keeps alive the tradition of the small town airport. It was a grass strip at the time, set in the rolling hills along the Susquehanna River, a very nostalgic place indeed. Older airplanes feel right at home there at the family run airport.

I paid only $8,000 dollars for N7203J and over the next year began the process of bringing her back to life. With a new engine and a paint job, she was well on her way to recovery. That was back in 1991 when I flew solely for pleasure and had only about 200 hours under my belt. N7203J would go on to teach me more about flying than any human being could ever expound from his wealth of personal experience. She was a very docile airplane, designed for training, and oh so forgiving of the many errors of a fledgling pilot. During the next six years 03 Juliet and I dodged fearsome thunderstorms, landed in blistering crosswinds, shared the beauty of full circle rainbows and summer sunsets, and began to navigate to destinations near and far. With that airplane I started to stretch my wings and gain the skill that can only come from experience. I took Dave on as a partner, and we installed a new interior and more modern avionics. Dave eventually bought me out completely, helping to finance my dream of a career in aviation. He moved to York, Pennsylvania, a short while after purchasing the airplane and joined a physicians' group that eventually made him a partner. Over time he continued the improvements and today he has about the nicest Piper Cherokee 140 I have ever seen. It's a real show piece, the kind of airplane you would see featured on the cover of an aviation magazine.(Two years after our trip it did in-fact end up on the cover of "Piper Owner Magazine." Reference issue number 281 August 2005) His attachment to 03 Juliet was founded in his own flying experiences with her, and she has become one of his most valued possessions. He had offered me his airplane and had entrusted it to me for a journey of some 2400 miles. I knew how much that aircraft meant to Dave, and I didn't take lightly the privilege he had bestowed upon me. The operating costs alone would run nearly $2,000 dollars, and he had donated its service free of charge. The only reward I could provide was my heartfelt thanks, for without Dave the program would have suffered a major setback, if it were able to continue at all. His only other more tangible benefit was that he and I would fly the return trip together, completing a flight that we had been pondering for years but were unable to manage due to our work schedules. This way he would only need a week off from work. It was a great trip back home, a memorable adventure in its own right.

Dave Speranza and myself with his airplane N7203J (Dave is the handsome guy on the left.)

The final member of our fundraising team, Reenie Johnson, signed on around the end of June to put her public relations expertise on our list of valuable assets. Reenie spent countless hours putting together press releases and making contacts on a local level as Beth continued to promote our program from the headquarters of The Leukemia and Lymphoma Society.

The team of Furman, Speranza, Millington, Fink, Glass, Johnson, and Rencavage worked hard during our spare time doing everything possible to raise funds. Kim and I did interviews with the local television stations and newspapers. I even did a twenty minute segment for public television in Central Pennsylvania to promote our flight and honor Kimberly's battle with leukemia. Captain Patrick Flannery contacted John Perkinson of the Airline Pilots Association who arranged for an article about our project to be published in its national magazine, reaching over 60,000 pilots and their families. USAirways also helped out with an article in The Hub Magazine, their corporate newsletter. We approached our employers and former employers and the representatives of companies that make the medicines involved in battling leukemia. We were searching hard for donations, petitioning many local businesses. Even the local Lions Club, North Pocono Minisink, took up our cause. I personally hate to ask for contributions, but it was for a good cause, one in which we all truly believed. That desire to help set aside any apprehensions I had. Over time, our fund raising skills grew more and more professional and our efforts began to reap rewards.

During the summer the donations started to roll in, ten dollars from here, a hundred dollars from there. People from all over the country were sending what they could afford and little by little we were making progress. Some of the donations came with notes of encouragement from families who had lost loved ones to cancer. One came from the wife of a USAirways captain who had lost her husband to bladder cancer, another from a USAirways mechanic who had battled leukemia himself and won. People that we had known and lost touch with over the years had seen the articles or television spots and sent donations with heart warming notes that let us know that friendship has no sense of time.

One woman I had worked with over six years prior to our trip sent a check from her and her husband for 500 dollars, a truly magnanimous gift. I don't mention their names here out of respect for their privacy, but I thank Dr. and Mrs. C. from the bottom of my heart. The Scrobola family, legends in aviation in our local area, donated 200 dollars. Mr. Bill Ferri of Ferri's Pizza pulled me aside as soon as I walked in the door and donated 100 dollars before I even had the chance to ask for a donation. He posted a newspaper article about our trip in his pizza shop which generated additional donations as well. Mr. Tom Pugh, the CEO of The John Heinz Institute, a rehab hospital in Wilkes-Barre, Pennsylvania, received our petition personally. When the corporate finances were too tight for a direct donation he put the word out to all of the department heads in the facility. The employees themselves got together and created a special basket raffle and fund raising event that brought in 1,300 dollars in donations from my former comrades.

In the end, we were blessed by donations from many wonderful people and even a few corporations. Some companies provided things that we would need for the trip. Avfuels donated 1,000 dollars for fuel, and Jeppesen provided charts and software for flight planning. Roche Pharmaceuticals overwhelmed us with their kindness by donating 1,500 dollars to Wings Across America, and putting their public relations people to work for us as well. They set up television spots with the local news and radio interviews that Kimberly did in the York area. I have to thank Mr. Gordon Wareham of Roche for the unending support he offered for no other reason than that he believed in us and what we were trying to accomplish. We did not receive as much support from the corporate world as we had anticipated. The vast majority of the funds we raised came from individuals as corporate sponsorship waned in the wake of an uninspired economy. I am grateful for those corporations that gave and for the individuals who did whatever they could, especially the employees of Allegheny Airlines, USAirways and The John Heinz Institute. They stepped up to the plate, even when finances had prohibited the support of their employers. To them I offer my most sincere thanks. Everyone who gave, family, friends, and acquaintances, especially those whom we never got to meet personally deserve our gratitude. I take great pride in the fact that every penny donated, with the exception of funds given for a specific purpose by the donor, (I.E. Fuel purchased with money from Avfuels) was given to The Leukemia and Lymphoma Society. Kim and I pitched in to cover our own expenses as did Dave Speranza who even went a step further covering all of the fuel expenses for the return trip and donating an extra amount of cash above and beyond the cost of operating his aircraft.

As it turned out, we received such a heartwarming response that the fund raising became as much a treasured memory as the flight itself.

We often find ourselves blaming God for the illnesses that befall us or those that we love. I don't have answers that apply to each and every situation with the exception of one. If each of us donated a portion of our time, talents, and finances to a cause that touched our hearts, we would most certainly see our efforts reflected in a world that was just a little brighter and a little more caring. We could close the gap between the way things are and the way things ought to be. We might even be able to build a world with fewer ailments and a little less suffering. The citizens of the United States give billions of dollars each year to charities; billions more are needed. I'd like to say thank you to those who have helped us add our drop to that bucket. With continued support from us all, a brighter future is no doubt on the way.

PART 3

I hope you never fear those moun-
tains in the distance
Never settle for the path of least resistance
Living might mean taking chances,
but they're worth taking
Lovin' might be a mistake but it's worth making
Don't let some hell bent heart leave you bitter
When you come close to selling out, Reconsider
Give the heavens above more than
just a passing glance
And when you get the choice to sit it out or dance
I hope you dance

Lee Ann Womac
"I Hope You Dance"

Chapter 5

It was September 20, 2003 and we were airborne. We had departed the Wilkes-Barre/ Scranton International Airport about an hour late, as we had waited out the early morning fog along our route of flight. The weather reports had indicated improving conditions to the east, but I had not seen any signs of such improvement. We motored along between cloud layers with occasional glimpses of the ground flicking by like aerial snapshots of Eastern Pennsylvania and Western New Jersey. Somewhere east of the Delaware Water Gap mother nature put on the lens cap and the snapshots stopped. We were aloft now in another world, a world of cloud and fog, aloft in the heavens with our only contact to reality being the voice of air traffic control and the needles and displays of our navigation equipment. I monitored the weather at a nearby airport on our number two radio and realized that those forecasted improvements had not yet arrived. The clouds went all the way down to just a few hundred feet above the ground, not a crisis by any means, yet certainly a cause for concern. There was only one real issue. Flying a single engine aircraft over terrain that was covered by low overcast skies is all well and good, until that engine fails. Although failures of modern aircraft engines are rare, they do occur. If that were to happen our flight would be cut short, and we would begin to glide, but glide to where? The earth itself had the appearance of an endless sea of soft white linen, but as Antoine de Saint Exupery once wrote, "But you want to remember that below the sea of clouds lies eternity." Without a visual sighting of a possible landing area our chances of a desirable outcome would be extremely limited. As we entered the clouds, our engine driven artificial horizon and directional gyro would become inaccurate, no longer having a source of vacuum to sustain their operations. I would be forced to handle an engine out emergency while keeping the airplane right side up by the use of the turn coordinator, an electrically powered gyro which is used as a back up device to the artificial horizon. That is not a position I would enjoy being in as it is one in which I would be starting to paint myself into a box, limiting my options and breaking one of aviation's most sacred rules, "Always have a way out." The weather was forecast to be considerably better than it was, hence my decision to depart. While I was contemplating a diversion to an alternate destination in order to sit and wait out the improvements, the weather to the east started to clear.

As we passed the Huguenot VOR (a radio navigation aid about 25 miles west of the Hudson River) the snapshots had returned. Further east, the world became a full color motion picture as the mist gave way to clearing skies, and those gentle rays of morning light illuminated the city of New York off in the distance.

An aerial view of the high striated cliffs of the Palisades

We joined the Hudson River just south of the Tapanzee Bridge and the visual feast had begun. We descended to just 800 feet to duck beneath the floor of the busy New York airspace and floated past the high striated cliffs of the Palisades as our shadow graced their peaks. We were low; it was a requirement to maintain our legality in this area, but it was fun as well. We were skimming along just a few hundred feet above the water in a corridor of airspace that let us enjoy a view of New York City few people experience but all would certainly appreciate.

The towers of the George Washington Bridge soon flashed beneath our wings. Ah, those poor souls stuck in that traffic beneath us. I've known their pain first hand.

The rising sun gave the city a surreal look, like that of an impressionist painting,

A little further south and the Empire State Building dominated the skyline off our left wing. The morning sun rising in the east combined with the remaining moisture and gave the city a surreal look, like that of an impressionist painting, a beautiful sight indeed. As we flew down that river, beneath the city's tallest buildings, I remember thinking that flight is certainly a gift to mankind. I know all about the physics that make it possible, yet it has never been anything less than magic to me on such a beautiful morning aloft. Kimberly was busy flashing photos of the most scenic flight on the entire east coast as we began a gradual descent to 500 feet to get a closer look at Lady Liberty.

I pointed out the sight of the World Trade Center, the sight where thousands of innocent people lost their lives at the hands of men who were jealous of the freedom and prosperity of our great nation. They hid their acts of hatred under the guise of religious beliefs. I cannot believe that any religion would propagate the mass murder committed on that fateful day. For a moment, I was saddened thinking of those who were lost and their loved ones left behind. We should never forget them or those who defend our way of life and prevent such tragedy from ruling our future.

Level now at 500 feet, just 104 feet above her torch, Lady Liberty loomed in the windshield. We spoke of what a spectacular sight she must have been to those who endured the long journey to the new world. Our grandmother was one of those brave, determined individuals. Mary Francis came to this country on a ship in 1914, at the tender young age of fourteen years. She was alone, save those few friends she made during her journey. She was greeted by that vision of hope in the harbor. The Statue of Liberty meant so much to those immigrants, signifying the end of one journey yet the beginning of another. Now, just two generations later, the grandchildren of that penniless immigrant flew over that very same symbol of freedom and opportunity.

Lady Liberty brought back memories of our grandmother.

Looking down at the Statue of Liberty, I was overwhelmed with feelings of patriotism. Looking back on our trip, as I wrote these words, I realized why. I looked back on that moment in flight and the thoughts of my grandmother. Because of her courage and the sacrifices made by her and our parents, we were given the opportunity to take our generation a step further and build a better life for ourselves and our children. Our country is filled with millions of similar stories. It is, in my opinion, the finest country in the world, bar none. It has its troubles, but as our cross country trip had proven, it is still a land of freedom and opportunity. It is still filled with hard-working, caring people, regardless of the visions of desperation we see when we focus our attention on those featured in the news and on television. The people they make television shows about are out there, but so are the Americans who work hard everyday and do their best to make a go of it on their own. We met them, a lot of them, during our trip. America ***is*** the land of the free and the home of the brave.

As our aerial tour of New York came to a close, we climbed to 1,000 feet and crossed over the Veranzano Bridge, putting New York's harbor behind us. We flew directly to the Colt's Neck VOR, then continued our climb as we turned to the west, facing our 2400 mile journey to San Diego. Our wings had passed over the Atlantic Seaboard, and we were finally headed in the direction of our destination.

With the sun at our backs, we passed over New Jersey and on into Pennsylvania once more. This time our view of Pennsylvania had changed from one of emerald fields and rolling hills to a vista of suburbia unleashed. As we flew north of the Philadelphia International Airport, the dominant feature of the landscape was that of urban sprawl. It's amazing how quickly things change. As far as the eye can see, from Philadelphia all the way north to Allentown, the farmlands have faded, receding to memory, and giving way to the growth of the city. It seems that every time I pass over this area there is a new warehouse or housing development under construction. Philadelphia and its suburbs contribute a great deal to the state, but they are a different Pennsylvania from the small coal town Kim and I grew up in or the rural countryside I now call my home. Suburbia soon gave way to the farmlands of Lancaster where the roads are shared with the horses and buggies of the Pennsylvania Dutch, a sharp contrast to the towering city skyline we had just left behind.

On approach to Harrisburg International Airport. Kim's make up was done, her checklist completed.

In a little under three hours from our departure, we were on approach to the Harrisburg International Airport. I turned to Kim and asked her to help look for traffic in preparation for landing while I completed the last few items on the checklist. My next glance at Kim took me by surprise. It seems our definitions of preparation for landing were a little different. She had decided to whip out her purse and put on her make up. We bobbled up and down in the mid-morning turbulence while Kim performed her touch ups in preparation for our interview with the local news. I pictured her ending up looking like a circus clown as her hand got jostled with every little bounce of our trusty Cherokee. Somehow she pulled it off, though, and when we finally unbuckled and stepped out on to the ramp at Midcoast Aviation, she was camera ready. It taught me a valuable lesson and from that point forward we had two parts to our before landing checklist, hers and mine. Hers came about ten minutes prior to every landing. I guess there were a few adjustments to be made when sharing an adventure with a member of the fairer sex.

The direct flight from Scranton to Harrisburg would take just under an hour, but our little scenic tour had extended that time considerably. It was all in the plan, though, a plan that included meeting up with Duane and my family once more at the Harrisburg airport and affording Duane an opportunity to share one last goodbye with Kimberly.

Kimberly and I finished up the interviews with Harrisburg's ABC affiliate and the York Dispatch, then headed inside. We had a regular greeting party and all we had done was fly the first leg of our trip. Our little rendezvous was capped off by a lunch in the airport restaurant. It was truly enjoyable, but we had to cut it short in the interest of making Terre Haute by sun down. Before I knew it, we were back in the airplane finishing up our checklists, ready to depart.

Just as I was picking up our departure clearance from the ground controller at Harrisburg, a friendly voice came over the radio wishing us good luck. It was captain Byron Barnes of Allegheny Airlines on his way to the runway himself. He and I had just flown together the previous week, and it was great to hear from him before we left. At the time of our flight, I was employed by Allegheny Airlines, a wholly owned subsidiary of USAirways. We flew thirty-seven passenger Dehavilland Dash8 aircraft throughout the Northeastern United States feeding the hub airports of Boston, LaGuardia, Philadelphia, Washington National, and Pittsburgh, some of the country's busiest. It was a great job, a job that allowed me the opportunity to fly with some great people, people who knew more about flying than I could ever hope to know. Names like Paul Detwiler, Mike McCambridge, Rick Flemming, Patrick Flannery,

Walter Fenstermacher, Lance Nothstein, Mark Henry and Mike Quinn will forever be remembered for their skill as pilots and their valor as men. The list could go on and on, and I offer my apologies to those I have not mentioned, but they know who they are and are appreciated as well by anyone who has had the opportunity to fly with them.

The spring of 2004 saw the end of Allegheny Airlines. Some of its people hung in there through the merger with its sister airline Piedmont. I left when I was given the opportunity to fly a new airplane, my first pure jet aircraft, the Embraer 170 for MidAtlantic Airlines, yet another division of USAirways. A good number of former Allegheny pilots went to MidAtlantic and are hoping for the best as USAirways works through its bankruptcy. I would like to honor all of the people whom I worked with at Allegheny Airlines who stood strong and maintained their integrity and professionalism through some very difficult times. I am glad to have had the chance to work with them.

Kimberly and I watched Byron and his crew depart runway 13 as we taxied. Our departure was pretty much by the book, checklist complete, cleared for takeoff and we were once again underway. We turned to the west and I decided it was far more prudent for me to pick up an instrument clearance and climb through the wall of cumulus build ups that we were fast approaching than to duck beneath them. The misty mountain tops became less and less inviting the closer we got. We kept our eyes peeled for cell towers and other obstructions as we bounced along above the ridges of the Allegheny Mountains and beneath the ragged bases of the early afternoon build ups. Before long, air traffic control came up with a clearance that allowed us to climb into the clouds without breaking any regulations.

If you can imagine the view from the inside of a ping pong ball you can pretty much understand what it looks like when flying inside of a cloud. You can see nothing but the aircraft itself and a grayish-white mist no matter in what direction you choose to look. The only salvation you have is the collection of instruments on your panel. They keep you upright, they tell you of your speed, your heading, your altitude, and your rate of climb or descent. Keep your scan going in a sort of relaxed concentration from one instrument to the next and everything stays in order. Keep climbing and eventually you'll break through the top of the layer, returning once more to the world of sight. When we broke out, somewhere around 5,000 feet, we were in a valley between two towering walls of cumulus clouds that formed like mountain peaks to our left and right. I cheated a few degrees from our assigned heading to keep us in the smooth clear air as we dodged those turbulent cumulus ridges and continued our quest for altitude. I've always liked to fly in these conditions, facing the challenge of providing the smoothest possible ride while gracefully banking around the mountains of rising air marked by the brilliant white clouds of a summer afternoon. Kimberly enjoyed the view as well and photographed the roiling white brilliance set against the azure sky.

We leveled at 8,000 feet in clear skies above the Allegheny Mountains and their images mimicked in the cumulus clouds that floated beneath us. Soon we were just south of the Pittsburgh International Airport, and the clouds had disappeared altogether.

I remember thinking that today I was among those little weekend warrior bug smashers we often referred to when flying for the airlines. I say that with great affection and respect because I was very grateful to be counted among those pilots who fly for the fun of it. The freedom of flight is a lure that draws many of us into the world of aviation, and I think that the pilot who is able to hold on to that sense of freedom more than any other is the private aircraft owner.

The private aircraft owner is a special breed of pilot for he possesses the romance of the early days of aviation. He chooses where he wants to go and when. He flies not for duty but for love, the pure love of flight that is enjoyed on those crystal clear summer evenings aloft over fields of emerald and gold and lakes like pools of liquid sunlight that shimmer in a gentle evening breeze. He flies for the challenge of the day, sliding down an invisible radio wave that provides safe passage through the blinding darkness of a rain cloud to the runway below. But most of all he flies because he chooses to fly. He chooses the day that he shall fly, the challenges that he shall face, and the destinations that he shall enjoy. That is the glory of being an aircraft owner. It is what I left behind in my pursuit of a career as a professional aviator. I do love so very many things about my chosen career, and it has made me a far better pilot than I could possibly have become on my own. Someday, however, I pray I will be able to add to my professional flying the joy that can only come from owning your own aircraft.

Leaving Pittsburgh behind, we crossed the Ohio River and officially entered the Midwest. The southeastern corner of Ohio looked relatively flat from our altitude, but as we got lower the terrain appeared much less inviting. The plain-like features were interrupted by deep crevices, as if when God rolled out the land he pressed it with large odd shaped cookie cutters, making it far less suitable for an emergency landing than it had originally appeared. We weren't over that terrain for long when we started looking for the Harrison County Airport, our next fuel stop. I over flew the field and did a teardrop entry into the downwind leg, touching down in the first third of the runway and taxiing off just moments later.

Every airport along our route had etched itself into our memories for one reason or another. Harrison County was no different. It was a great little airport with certain idiosyncrasies that made it special, like having to pump your own fuel, a labor of love if truth be told. I kind of enjoyed taking care of the little airplane that was working so hard for us that day.

The real memory-maker though was one of the gentlemen that was sitting with a group of hanger flyers gathered in the office talking away the day as afternoon turned to evening. I won't mention his name here, but Kim and I referred to him as "Tall Tale" for the rest of the trip whenever he came up in conversation. Now granted I've always felt that hanger flying had origins in the fish tales told by those intrepid souls who spent their days at sea and their evenings in the local pubs exaggerating their exploits. In the sixteen years since I started flying I've spent my share of warm summer evenings hanging around the local airport after a particularly enjoyable flight, listening and talking with my friends about our adventures aloft. The difference with old "Tall Tale" was that he hadn't developed the skill of assessing his audience. When he queried us as to our destination, we told him of our flight and the time it would take us to complete it. Old "Tall Tale" then decided to regale us with his experiences on a similar flight that he had taken the previous year. Now, supposedly, he just got up one morning and decided he would fly to the west coast, to San Diego in fact. He claimed he just jumped in his airplane, a model with not much better performance than old 03 Juliet, and flew there in under ten hours, no planning or preparation required, just flew until the gas got a little low then refueled and flew on.

No harm, no foul, is truly my take on a discussion like this, but why spin such a yarn to the people who just got done telling you that they knew how long it would take and what they had done to prepare for such an odyssey. A liar is bad, but a bad liar is even worse. Of course, old "Tall Tale" was exaggerating. If he had ever even completed such a flight, it would be impossible to do it in the time he proposed with the aircraft he was flying, but I chose the higher road and thanked him for his tidbits of advice as we made our exit. He was still reeling in his compatriots when we closed the door behind us.

I had originally planned to let Kimberly perform the take off at Harrison County, but an evaluation of the field and the little challenges it presented wouldn't allow for that. The Airport Facilities Directory is a little green book that provides unfamiliar pilots with information such as field elevation, obstructions, traffic pattern altitudes and things of that nature. Our AFD was telling us that Harrison County had nearly a 2 percent incline to overcome when departing to the Northwest. It might not sound like much, but it would have a considerable impact on the performance of our little Cherokee. The five knot headwind favored the Northwest departure, but just slightly. A southeast departure would put us in a tailwind situation heading toward an embankment off the departure end of the runway. It would be a considerable obstacle to clear, especially with the wind at our backs. It was a judgment call, no doubt, and I opted for the uphill departure. I decided that I would perform a maximum performance take off tempered with a rule of thumb evaluation. A book I had read some time ago had stated that if you hadn't achieved 75 percent of your required flying speed by the halfway point on the runway it was time to abort.

I discussed the situation with Kim, to keep her in the loop, and we made our final radio call as I positioned us on the runway. I held the brakes and smoothly advanced the throttle toward the firewall. One quick glance at the instruments showed that we were creating maximum power and that all of the engine instruments were hovering comfortably in the green.

As I let go of the brakes, we watched together while the little Cherokee began to build airspeed, slowly at first, then a little faster. I pulled back on the stick right at rotation speed, and we were in the air by the middle of the runway, no sweat! As the airspeed increased, I pitched up and set the airplane for its best angle of climb, giving it the greatest gain in altitude over a given distance. The propeller clawed at the evening air and before long we found ourselves safely airborne once again, retracting the flaps and completing the climb checklist at the leisurely pace we had become accustomed to.

The fields ahead were perfect, flat and low, with boundary lines that ran North-South and East-West. Just pick a road aligned with a boundary line of your required direction and in our case, head toward the setting sun. We flew low on this leg, just 500 to 1,000 feet above the ground at times. Compared with the rolling hills of the Northeast, the Midwest farmlands are a pilot's paradise. You still have to pay attention to towers and power lines, but, for the most part, it's emergency landing sites as far as the eye can see. That added margin of comfort brings a great deal of enjoyment to the flight.

Somewhere around the border of Ohio and Indiana Kimberly noticed a field that had been turned into crop art. It could only be fully appreciated by those of us who were lucky enough to stumble upon it from the air. The entire field was a huge caricature of man running. The title across the top read "Harvest Land." It was really cool to see, and I just can't imagine how they did such detailed work without seeing what it was they were creating. We both enjoyed it, and, of course the team photographer grabbed a few snapshots for memory's sake. Hats off to the artist, or was it really done by aliens. (Insert suspenseful music of your choice.)

The next photo is of the crop art. The original showed it rather distinctly but this is the best I was able to do with photo shop. It really was quite detailed and very cool! It's in the lighter field in front of the wing on the next page.(Utilize the e-book zoom for greater detail)

We relished this leg of the flight, watching the cities of Columbus, Dayton and Indianapolis glide by to the left and right of our course. Up to this point in our trip, we spent a great deal of our time aloft tending to the business of flying. We departed under instrument conditions and then headed right off to carefully navigate the busy airspace of New York and Philadelphia. Then we were in the clouds again on our way out of Harrisburg, followed by a slightly tricky departure from

Harrison County. It had been a pretty busy day, a day marked by flights that were more laborious than the usual little hour long hops that Kim had accompanied me on in the past. She had never spent any serious time in a light airplane prior to our trip, and I was a little concerned as to how she would handle it. She did great. She jumped right in, handing me charts and keeping the cockpit organized for me. She even answered a few of the radio calls. Now, in this less congested airspace, I gave her the controls for a while. She was a natural, holding her heading and maintaining her altitude as though she had been flying for years. She would certainly have made some of my former students jealous if they saw what an easy time she was having her very first time at the controls. After about ten or fifteen minutes, the thrill became a chore, and she handed the controls back to me. We laughed and joked about our hectic day as we navigated over the farmlands in true barnstormer fashion, low and slow in the smooth calm air of a summer evening.

The aerial sunset was spectacular. That golden disc faded to red as it sank softly into the distant horizon, leaving in its wake an explosion of color in endless shades of red, orange and magenta. The age old saying applies to flyers as well as sailors: "Red sky at night sailors delight." Tomorrow holds promise.

.

A beautiful end to a great first day. Red sky at night..........

Twilight became darkness and the farmhouses shimmered like beacons of humanity scattered across the vast black plains. We reveled in the beauty of night flight and the feeling of peaceful isolation, bathed in the glow of the moon and the twinkle of the evening stars.

Soon, Terre Haute was in our sights and we followed the alternating flashes of white and green light to the Terre Haute International Airport. We were tired and hungry after nearly eight hours in the air. We were closing quickly on the sixteen hour mark for the day counting our preparations and occasional stops. We rallied ourselves for one final landing as I aligned us with runway 32 while backing myself up with the instrument landing system frequency, just in case. Tired eyes can sometimes make mistakes. I followed those radio waves right down to the runway threshold, then pulled the power to idle and let old 03Juliet settle into the landing with a touch of backpressure and a stereophonic chirp of rubber on pavement. I lowered the nose wheel gently and rolled to a stop, clearing the runway while the ground controller gave us taxi instructions to the Terre Haute Air Center and bid us good night.

Once on the ramp we were greeted by a lineman named Craig he was the embodiment of the Midwestern work ethic. He literally laid down a small red carpet for us as we gathered our belongings and headed toward the FBO. (FBO is an abbreviation for Fixed Base Operator. It can be likened to a service station of sorts for airplanes. The personnel are specially trained in the skills of handling and refueling aircraft of all types ranging from airliners to business jets to little Cherokees like 03 Juliet.) Crystal the customer service representative tended to our need for directions and provided transportation to the Comfort Inn where we would spend the night. We were far too tired and unfamiliar with the local area to search out a nice place to eat. Instead, our first day of travel would be capped off by a stop at Burger King and a few phone calls home. It would turn out to be the longest day of the trip and fulfilled the words of the Robert Frost quote Doug had chosen for our website:

"...But I have promises to keep, and miles to go before I sleep..."

Chapter 6

The Comfort Inn, and more importantly its general manager, was very gracious to us. I distinctly remember my phone call to him. He didn't hesitate for a second when I told him of our endeavor and asked for his support. He was the voice of kindness in a sea of rejection. To my surprise, not very many hotels were willing to offer a night's stay to help our cause. Only Mister Sanders in Terre Haute and a young lady at the Comfort Inn in Tucson were generous enough to help out. Unfortunately, we never did get the chance to overnight in Tucson as weather delays had changed our plans. So, as it turned out, our evening with Mr. Sanders and his staff was the only free night's stay of the trip. I wanted to thank him personally before we left that morning, and I was glad that we had taken the time to do so. He was just as kind and pleasant in person as he had been on the phone. Kimberly said it best, "He was a big man with an even bigger heart, a true credit to the organization he worked for." He was pleasantly attending to the needs of another customer as we waved goodbye and finished packing our things into the blazer.

When we returned to the Terre Haute Air Center, Crystal was again working the desk. Kimberly spoke with her about our trip, and I could hear the enthusiasm in her voice when she spoke of our first day of flight. It was refreshing to hear that she was so into our adventure. I had secretly held a concern that she wouldn't enjoy the flying once we actually spent some serious time aloft. Her smile and fervor while speaking with Crystal dispelled that worry. In the preliminary planning I had decided to file an instrument flight plan for our flight into St. Louis. It would reduce our nuisance factor to the air traffic controllers as we traversed their busy airspace. My original flight plan called for our next stop to be the Spirit of St. Louis Airport just west of the city. While that would be a great airport at which to stop, I decided that the aerial tour of our country could not possibly be complete without a photo of The Gateway Arch from the air. Cahokia, Illinois, is just across the mighty Mississippi River from St. Louis. More importantly, it's just east of The Gateway Arch. I was looking over the charts the night before as Kim was preparing for bed when I was struck with this geographic realization. Because of my decision to change our arrival airport, I found myself in need of the approach plates for St. Louis Downtown Airport. Crystal came to our rescue, and we were back on track in no time.

After the rather long day we had put in on Saturday, we decided to adopt a more leisurely pace. Besides the adventure was in the trip, not the destination. While I was getting the airplane topped off with fuel and adding a quart of oil, Kimberly sat on the wing in the mid-morning sun. I knew for certain she was enjoying herself. She looked so at ease sitting Indian style and writing in the journal I had given her. Her life is usually so hectic, trying to keep up with the ever-changing world of medicine, that she never has the chance to enjoy such things as leisure reading or journal writing. She was caught up in our little project, relaxed and at ease from the moment our trip had started. It gave her a sort of tranquility that shined through in her every action.

I think that once we strip away the stresses of everyday life and have the chance to truly relax, we get a glimpse of who it is that we have become. Kimberly is almost always a happy and positive individual, and as the trip progressed those characteristics were continually enhanced until the woman whom I flew with became the charming little girl I had always remembered my sister to be. She had suffered and struggled so much over recent years, and although she persevered and always put forth her best, I seemed to sense her struggle. She was better now, and it was evident in a lot of ways.

Kimberly writing in her journal on the wing of 03J.

It wasn't long until we were airborne again, enroute to St. Louis Downtown Airport in Cahokia, Illinois. The landscape that we had watched fade to darkness the evening before had returned, accentuated by clear blue skies and unlimited visibility. Looking down upon the fields of southern Illinois, I was suddenly reminded that this is barnstorming country.

At one time barnstorming was a great way for a pilot to make a living. After World War I a pilot could pick up a used Curtis Jenny for somewhere in the neighborhood of two to three hundred dollars. With aviation in its infancy, and a great number of Americans enamored with the mysteries of flight, an ambitious young aviator could earn upwards of 500 dollars a day hopping rides out of hayfields.

In my mind's eye I saw through the boundaries of time. There beneath us, just outside that Midwestern town, was a long rectangular hayfield. It was perfect, set between the baseball park and the lake. Looking down I could envision him, the barnstormer, shattering the silence of the morning with his bright red biplane. As he crossed over the sparkling lake, his wings would glisten, winking rays of sunlight at those whose eyes, and hearts I might add, had followed their ears. The sound would fade to the whistle of wind through the flying wires as he pulled the power to idle and glided to the ground with a gentle thud. Then came the muffled rumble of tire on grass as he slowed to a stop, and with a sudden burst of power, he would taxi off to the side of the field. He would, of course, need the owner's permission to operate out of that field, and it would come with no more payment than the promise of an adventure in the air. It was an effective barter that left both parties equally satisfied. Shortly afterward he would uncover his front cockpit. Putting aside his supplies, he would pull out his sign, and as soon as the crowd had arrived, the sales pitch would begin. "See your house from the air! Ten minutes in the heavens where only the birds and the angels have gone before! For just five dollars folks, you can enjoy the experience of a lifetime!" The people of that town would gather in awe, gladly spending their hard earned money for their chance to break free of the bonds of gravity and view the world from on high.

It was a romantic period in the history of flight,

a time before rules and regulations, a time when a pilot's well-being was governed solely by his abilities. The flights themselves were spectacular. Out in the breeze of the open cockpit, you could feel the wind and touch the sky. They were low altitude jaunts around your hometown, a chance to see from the air the places you had known all of your life: your home, your church or your schoolyard. These pilots provided the true experience of flight, the flight that is held in the dreams of a child, low and slow and filled with the magic of a first experience in the air.

At the end of the day the barnstormer would be treated as an honored guest among the people of the town. They were all too eager to hear of his adventurous life. He would often spend his nights beneath his wing, basking in the glow of the moon and the starlight of a Midwestern sky.

I've looked through time to capture the essence of the life of a 1920's sky gypsy. My time travel did not require the exotic science of a time machine. Instead it was done in the only true manner of time travel I have come to know. It is a method available to each and every one of us, a simple process. Just open a book and embrace its contents. I've read about barnstormers in dozens of books over the years. Each one allowed me a chance to gaze upon the lives they led so many years before I was born. Flying over the state of Illinois on a warm September morning, it was those literary visits to the 1920's that dominated my thoughts.

Before long the St. Louis skyline rose into view. It all came into a sort of gradual focus, until the gateway to the west appeared on the far shore of the mighty Mississippi River. I requested a wide traffic pattern as we positioned ourselves for landing on runway 12 right. The controller granted my request, clearing me for the arch view approach. He knew what I was up to. It was another fantastic view. The huge metal arch threw reflections of sunbeams in our direction, welcoming us to St. Louis. Kimberly got some truly great photographs and there was a sense of excitement in the air. The Mississippi River and the city of St. Louis were definitive landmarks attesting to our progress.

The Mississippi and the view on the arch view approach into Cahokia.

Our plan was to eat lunch somewhere nearby, maybe get a few pictures of the arch from the ground and then press on, ending our day in Tulsa. It wasn't long after we set foot on the ramp that we were looking into alternate plans. Thunderstorms were booming all along our projected route. A cold front was making its way across our path, forcing the warm moist air in front of it upwards and filling the sky with a line of Midwestern thunderstorms that our little Cherokee had no business doing battle with. We decided to extend our sight-seeing tour of St. Louis and allow the storms to pass. We buttoned up 03 Juliet, installing its pitot cover and cowl plugs, then arranged for it to be put into a hanger for the night. A little extra TLC never hurts.

The FBO gave us a ride to the Holiday Inn Express and made sure we received the discounted rate they had arranged for us. This was a major discovery of the trip. The people at the FBOs seemed to be able negotiate a far better rate than we were able to arrange ourselves, no matter what type of discount or club memberships we tried to use. That was the case all across the country, and it saved us quite a bit of money throughout the course of the trip. We were only in the hotel room long enough to unload our gear. Then we grabbed a cab and headed to downtown St. Louis and The Gateway Arch. While crossing the bridge into St. Louis the view of the arch is inspiring; I thought its history was interesting as well.

Eero Saarinen competed in a design competition held during the years 1947 and 1948. The winner would have the opportunity to build the monument that celebrated the spirit of the western pioneer. Mr. Saarinen's magnificent design was the winner. The construction of the arch began in 1961; it was topped out in 1965, and dedicated in 1966. Unfortunately, the famed architect died in 1961, before having the opportunity to witness its completion. It is our nation's tallest monument rising to 630 feet from the banks of the Mississippi River. The great catenary curve that defines the St. Louis skyline remembers the spirit of the western pioneer and their willingness to risk it all in hopes of a brighter future.

We roamed the grounds of the Jefferson National Expansion Memorial Park where people were picnicking on the lawn. We shopped at the Levee Mercantile, an 1870's general store in the base of the arch. Unfortunately, the prices were more of the 2003 era. Then we took the tram ride to the top of the arch. It was an engineering feat in itself just to design the tram system. The tram cars seat up to five people in a small circular car that is completely enclosed, definitely not designed for those who have problems with claustrophobia. The ride itself, though, is relatively short and the view from the top is certainly worth the trip. Out the east-facing windows is the Mississippi River and the west-facing windows provide a pretty spectacular view of the city, highlighted by the old courthouse. That courthouse was the gathering point for wagon trains heading west. We crisscrossed the route that some of those wagon trains traversed. Each time we saw a sign or marker of their course I thought of what a truly treacherous journey those pioneers made in an effort to improve their lives. It all seemed pretty courageous to me. (No one can say Americans lack heart.) Our trip to the arch was a great sight seeing adventure that we enjoyed in the Rencavage family tradition of been there, done that, got the picture.

Speaking of pictures, Kim was inspired. We continued our tour of the city through the eyes of my photo journalist sister. She snapped photos everywhere we went. Some of them were nothing less than artistic. She truly has an eye for capturing the spirit of the moment. Because of her love of photography, I can relive our trip at any time I wish by merely gazing through the
more than 400 photographs she took during our trek westward.

The little girl that I had mentioned earlier was the sister I was traveling with now. She was so positive, so full of life, that she cast a sense of joy over everything we did. We were relaxed and settled into the sight-seeing mode, seeking out the secrets of the city. We were both enjoying our time together and the experience of meeting new people from different parts of the country, but I was starting to get hungry and finding a decent restaurant was beginning to dominate my thoughts.

When I was little I used to watch "The Incredible Hulk" starring Bill Bixby and Lou Ferrigno. The most memorable line of the show was frequently delivered by Bill Bixby, "Don't make me angry; you wouldn't like me when I'm angry." He would utter those words every few episodes, and they've stuck in my mind to this day. Well, as I've gotten older I've jokingly referred to that dialogue, or at least my own variant of it, based on my family's observations. My line is more like this: "Don't make me hungry; you wouldn't like me when I'm hungry." It seems, as my family has pointed out on occasion, I do have the tendency to get a little grumpy when my stomach starts to growl. My sister's husband Duane had mentioned the same trait in her as well. It must have come from being spoiled as a child with an overabundance of home-cooked meals prepared at a moment's notice by our mother. Anyway, it was in our best interest to get to a restaurant in the near future.

We chose Hannegan's Restaurant in the historical riverfront district. It was an excellent choice. We enjoyed our first full course dining experience of the trip amongst the lamp lit cobblestone streets of old St. Louis. It was the perfect finish to a long day of sightseeing.

View from the top of the arch!

My photojournalist sister at work.

The cobblestone streets of Old St.Louis.

Kim made friends everywhere she went.

Chapter 7

We were up early, anxious to do our best to make up for the time we had lost to weather the day before. The storms had passed, leaving in their wake low clouds and limited visibility which we hoped would dissipate as the morning progressed. We decided to grab breakfast and head over to the airport early. It would allow us to be in position, ready to resume our trip as soon as the weather lifted. The ceilings along our entire route to Tulsa were only 200 feet with just ½ mile of visibility. It was forecast to get better, but our quarter hour checks of the weather computer showed no improvement whatsoever. We were getting bored and a little anxious to get going, but the weather just kept getting worse. Around 11 o'clock in the morning the ceilings around St. Louis were rising; they were up to almost 1,000 feet. The weather along our route, however, was getting worse. The low ceilings and visibilities continued, and some popcorn thunderstorms were beginning to develop around seventy miles northeast of Tulsa. With all that moisture in the air, and the afternoon sun approaching fast, the thunderstorm situation would most likely get worse. It was time to go to plan B.

There was another front approaching, and the air in the northern end of it was much drier than that to the southwest. I decided then that we would proceed west-northwest to Kansas City. My decision would take us through the drier area of the front, and we would be in clear air within an hour. It would add some flying time to the trip, but it would get us back in the air and making progress a lot sooner than if we sat and waited until the route to Tulsa was passable for the little Cherokee with only one engine and no radar.

We said our good byes to the group of collegiate aviators who were waiting with us for weather improvements. They were hoping to complete their final leg back to campus. For Kimberly and me the time had come, and with a copy of the flight plan and weather in hand, we were again mounted up and heading west in the spirit of Lewis and Clark.

The air was pretty rough in the clouds with just an occasional break here and there. By the time we were forty miles west of St. Louis, though, the clouds dissipated. The ride, however, was not yet smooth, no matter what altitude we flew at. It was more of a light chop with occasional crests of moderate turbulence thrown in just to keep us awake. We fought our way across the entire state of Missouri, eventually choosing to stay as low as possible. The altitude would not help the turbulence, but staying low eliminated some of the headwind. Headwind had certainly become a consideration. At 8,000 feet it was 50 kts. and was cutting our groundspeed in half. Even at our chosen lower altitude, we were crossing the ground just under 70 mph at times. We watched in dismay as the traffic on the interstate below passed our little airplane caught in a battle for progress against a formidable wind.

Kansas City, Missouri, soon graced the horizon ahead, shrouded in sunshine and just a few higher level clouds. The turbulence continued, slamming us into our seat belts then relaxing and moments later repeating the cycle. I checked the ATIS (Automated Terminal Information System-A recorded briefing of the weather and any important information pertaining to the airport). It informed us of a quartering crosswind with gusts up to 30 kts (about 35 mph). Now 30 kts of wind on landing is worth paying attention to. When it is coming across the runway and trying to push your little airplane into the grass or the adjacent taxiway, it has to be dealt with. That in mind, we decided to forgo Kimberly's cosmetic before landing checklist and stick strictly to business.

When they cleared me for the visual approach to runway 1 right, we were still east of the field. I remember looking at the correlation between the tall concrete buildings of the city and the runway itself. Normally in such a small airplane I would stay inside the city and make a tight approach to the runway. That seemed impossible from my perspective. The city was just way too close to the airport. So we bobbled around a little more, circling the southern portion of the skyline to complete a base leg and final approach to the runway. On our way down the final approach course, I kept thinking to myself; "Wow TWA used to land 707's here!" We were on a path that put the tall buildings of the city off our one wing and a multitude of towers off the other, not much room for error.

We stayed on the extended centerline by crabbing into the wind, allowing the airplane's nose to point toward the wind in order to maintain our track across the ground. The conditions were constantly changing and I jockeyed the controls to accommodate them. While I worked the aileron and rudder pedals for the desired effect, I glanced over at Kim to see how she was handling all of this. I'm not sure if she was scared or not, but it was awfully quiet in the cockpit after we received our clearance to land. Crabbing can seem pretty confusing to someone who doesn't fly a lot. The nose of the airplane is not pointing toward the runway and it feels as if you are flying sideways. When the winds are as strong as they were that day, a good deal more of a crab is required and the sensation is enhanced. I held the crab and some additional airspeed to just about the point of touchdown. Then, as smoothly as possible, I rolled in some left aileron to keep our track and pressed on the right rudder to keep the airplane from turning. The result was a gentle kiss of the left main wheel on the pavement, followed by the almost imperceptible touching of the right main and finally the nose wheel. I reached for the brake with my right hand and Kimberly just smiled. We had faced our toughest landing challenge of the trip and come out of it with the best landing thus far. Kansas City here we come!

We taxied to the Executive Beechcraft ramp, pulling up between a multimillion dollar corporate jet and a DC-9 owned by Airtran airlines. There we sat in our little Cherokee, lost among the kind of airplanes that spend more on a single fuel stop than we would spend on our entire trip. Yes, we were still a customer, but definitely small potatoes when it all came down to dollars and cents. That didn't seem to matter to our hosts. By the time the propeller stopped spinning and we opened our door, a cheerful liaison was waiting at our wingtip in a courtesy van to take us to lunch.

We were hungry and starting to get tired. We had endured the constant battering of air turbulence for over two hours, and we needed a break. The driver took us on a quick tour of the city allowing us to photograph some of the many fountains Kansas City is noted for, and we discussed an alternate plan for the remainder of our day. We would again slow things up a bit and fly just one more leg with an overnight in Wichita. Things would have been different had it been a race, but it wasn't. It was an aerial tour of the country that we decided to enjoy and spread out a little more. Besides, there was a hurricane brewing over Mexico on a collision course with the final leg of our flight to San Diego. By slowing our pace, we felt we just might be able to let it pass and sneak in behind it to our destination. Wichita was not quite Hobbs, New Mexico, where we had originally hoped to end up that evening. I say that referring solely to distance covered as Hobbs has a long way to go in order to compete with Wichita. It was a good decision that put us in the right frame of mind to sit down and enjoy a delicious steak for lunch. It had to be steak, of course. You know the old saying; "When in Rome..."

By late afternoon we were back in the air. The clouds had dissipated and our only adversary was the wind. We trundled across the state of Kansas in continuous turbulence fighting the headwinds for progress. The ride was slow, but as the evening approached the sea of air that surrounded our little vessel had begun to calm. We looked around at the endless fields of summer, long flat fields extending off to the horizon, as green and bountiful as we had imagined our nation's heartland would be.

As we approached Wichita's Mid-continent Airport, a Stealth Fighter was departing the nearby air force base. He sped in front of us, screaming up to altitude in a long slow turn that culminated in a pair of distant contrails. It was our own private airshow. We spotted another aircraft as well, a Cessna 172 off to the northwest of our position. He appeared to be on a collision course with us. I queried air traffic control and they concurred. He was at our altitude and if we didn't take action a collision was imminent. We descended and watched as he passed unknowingly just a few hundred feet above our heads. It was nothing to be concerned about as we had caught him in our sights and avoided him with ample separation. It is worth mentioning because, although it is a big sky, it certainly pays to be attentive. I decided to try a short field approach to landing, just to brush up on the technique. Full flaps and right on speed, we cut the power and settled in for the landing. I got on the brakes just a bit and we easily cleared the runway at the first taxiway. The Piper Cherokee is one well-designed little airplane. In just one day it had seen us through an instrument departure, fierce headwinds, continual turbulence, and a crosswind landing that would have sent lesser airplanes to an alternate destination. It can take off short and land in just a few hundred feet. It really does a wonderful job thanks to those engineers at Piper back in the 60's.

We taxied for what seemed like an awfully long

time, then finally stopped at the far side of the field where our airplane would spend the night under the watchful eye of the staff at Executive Aircraft. Welcome to Wichita, Kansas, middle of the continent as they say, home of the Wichita Woo and Kim's alma mater Wichita State University.

The folks at Executive Aircraft arranged a tremendous deal for us with the Sheraton Four Points Hotel. We dined there and spent the night in the lap of luxury. Another day down, a long and bumpy day of limited progress, but a day of progress nonetheless.

Chapter 8

It was Tuesday morning and we were flying over Kansas, heading south now to Amarillo and back toward the course we had originally planned. Unfortunately, the wind was coming from the south. We would endure yet another day of slow groundspeeds and limited progress for the time we spent aloft.

The terrain was changing quickly, becoming more and more desolate with every mile. It was a quiet but intriguing desolation as the flat lands below had begun to show signs of character, uprisings and crevices that were sort of a precursor to the mountainous terrain of our future flights.

From southern Kansas across the Oklahoma panhandle and on into Texas, we didn't see a single sign of human life. There wasn't another airplane, not even a car, just mile after mile of oil wells pumping rhythmically in the breeze and the occasional windmill farm that would turn that breeze into electricity. There were roads that led to the oil wells, but they were no doubt used purely for the purpose of maintenance and not a soul was driving them that morning. We had settled in for the three hour leg to Amarillo, listening to nothing but the smooth constant rumble of the little Lycoming engine and the occasional frequency change assignment from air traffic control. It was then that Kimberly and I got engrossed in a conversation about our childhood.

Nothing but windmills and oil rigs from Oklahoma to Texas.

Now I've read plenty of books that discuss the years of torment that had brought an individual from childlike innocence into adulthood. Those stories, although often works of literature that are far beyond my abilities as writer, are usually very depressing tales of the devastation of their broken home or the tortures of life as an abused child. Fortunately for Kimberly and myself, no such darkness ever dominated our youth.

We were the product of two very loving parents and an extended family that was equally affectionate and caring. I'm not saying that we never faced trials and tribulations; that would simply not be true. What I am saying is that the love and understanding that permeated our family saw us through the tougher times and impressed upon our memories a certain sense that we were always blessed. Kim had started the conversation by stating how proud our father would have been of every member of the family. None of us has achieved tremendous financial success, yet we are all gainfully employed in a profession that we love. We do our best everyday to hold true to the principals and beliefs of our parents. They taught us the importance of things like love, respect and compassion toward others. Those values exist in our lives for one reason and one reason only. Our parents held true to those values themselves. They always put the needs of their children ahead of any personal agenda, or in their case, even personal need. I cannot remember my mother ever spending frivolously on herself, not once. My father would fall ill from time to time and at one point was out of work for quite a while, but my mother's diligence and loving support never wavered. In fact, because of her, we hadn't realized exactly how poor we were. It wasn't until years later that she disclosed all that she had been through.

Her family and friends should also be mentioned. My mother once told us of a time when she did not have any money to spare for Christmas presents. My dad was out of work and things had become increasingly difficult. Feeding and clothing her four children was absorbing her every resource. She cried to herself when she had returned home from the hospital after visiting my dad. She couldn't bear the thought of disappointing her four young children on Christmas morning. I know we would have understood, at least the older ones, but it broke her heart just to think about it. It was then that her friends and family came through like Santa Claus. The Charettes and Abilies and each of her brothers and sisters put together a wonderful Christmas for us. I was never aware of all that they had done, but after hearing the story I will never forget it. Kimberly and I spoke of the love that surrounded us all of our lives. It was there when we lost our father, and it became a source of strength during Kim's battle with leukemia.

Kim went on to bring up a few amusing stories from our past. She's such a caring and sensitive individual that she still felt guilty for youthful indiscretions that I myself had forgotten, but now find quite amusing. She apologized for an instance in our kitchen when I was eleven and she was twelve. We both had friends staying over and Kimberly had decided to show off her newly developed skill of making French toast for breakfast. First she cracked the eggs into the bowl, then picked up the wooden spoon to stir her batter. It was then that she found herself caught in a moment of weakness; a temptation struck her that she just could not resist. Her attention was diverted to the back of her younger brother's head. The spoon whistled through the air as I sat in disbelieving silence. It stopped suddenly with the loud crack of wood against my cranium and sent me whimpering off to my room. She continued with a story of the time she was assigned to awaken me for school and decided to comply by folding me up in the sofa bed I was sleeping on so peacefully.

These may sound like my personal versions of these tales, but in truth they were the guilty confessions of my sister given to me with apologies that day in the air over Texas, the guilt no doubt a reflection of our Catholic upbringing. (We have both since changed to different forms of Christianity.) I laughed out loud when Kim was finished and then did a bit of confessing myself. The truth of the matter was that on the fateful day of the spoon incident, I had been harassing Kim relentlessly in front of her friend. I was deserving of that smack and maybe a little more. The sofa episode was no different. She must have attempted to wake me up that day a half dozen times or more. Each time she tried and each time I ignored her as her frustration grew. The idea of folding me into a sofa bed was merely the solution to my ignorance, and a pretty creative one I might add.

Kimberly did have her fair share of adventures in misbehavior; we all did. It was to be expected in a family our size. Besides it gave our parents something to do, keeping us in line. Oh and by the way, I'm sure I'll pay for that last statement. After all I have two boys of my own to chase.

Confessions complete, we landed in Amarillo with a strong headwind and a growing need for lunch. The stop was uneventful and as I waited patiently for fuel and oil, Kimberly picked up some sandwiches at the restaurant next door. We finally refueled the Cherokee and our stomachs then left Amarillo in a slow climb to altitude. Our climb rate was inhibited considerably by the decreasing air density of the hot Texas afternoon and the elevation of our point of departure.

Back into the wind and the bumps we went, bobbling along over Lubbock and on into Hobbs, New Mexico. My mind, however, was looking ahead, focusing on our next big challenge, the Rocky Mountains. When we landed in Hobbs, I consulted with the local flight service station and our dispatcher friend, John Millington, back in Pennsylvania. It was clear that the strong winds and developing clouds would make our trek through the Guadelupe Pass more of an adventure than I was willing to put Kimberly through or myself, for that matter.

By the time I had decided that we would stay the night in Hobbs, Kim had arranged for a ride with the woman who was working the desk at the

FBO. Johanna was the delight of Hobbs, New Mexico. Kimberly and Johanna had been discussing everything from her illness to the internet. They became instant friends and Johanna was kind enough to take us to the hotel on her way home from work.

Johanna the delight of Hobbs ,New Mexico.

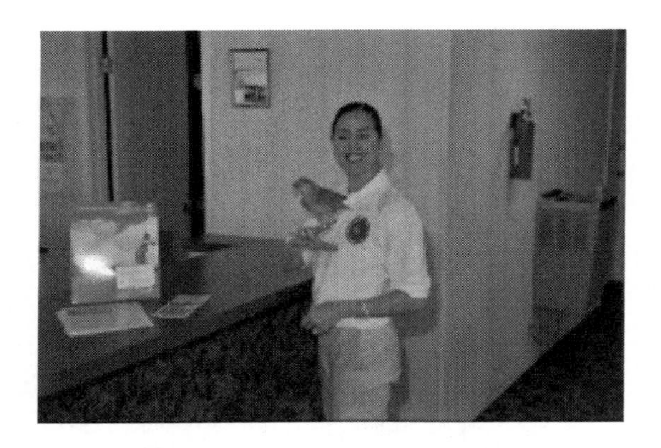

Kimberly with Mandy, the parrot mascot of Flower Aviation.

Hobbs was chosen as a stop because of its distance from the pass. It would allow us sufficient time to climb to a safe altitude for crossing the Rockies. We would be facing off with a continual decrease in climb performance all the way up to our crossing altitude. It was a concern that had to be addressed and stopping at Hobbs was the solution. It was far enough away from the pass to allow for the climb, but close enough to prevent fuel from being an additional problem.

Hobbs held very little merit as a city of interest,

but our stop there turned out to be an experience in itself. It is best described as a pipeline town, and the pipeline workers dominated the hotels in the area, including the one in which we were staying. Our hotel was really more of a two story motel, a home of the moment for the transient pipeline workers in pursuit of an honest day's work. They were a pretty diverse group, and I spent some time hanging out with Bob, an African American man, who worked the pipelines for a living. He was grilling on the second story walkway just outside our room. The aroma of chargrilled chicken accompanied his tales of life on the road and the sacrifices he made in the interest of earning a decent wage. His stories all seemed to begin with a smile, and he laughed when he told them. He really liked the work he did and the camaraderie he shared when he was out on the road, but it was tempered by his feelings of loneliness and his desire to spend more time with his family. I could relate to the words he spoke; pilots share in a similar conflict. I enjoyed my conversation with Bob while Kim spoke with Duane on her cell phone in our room. When she finished, I wished Bob a good night, and he returned my wishes with a smile and another hearty chuckle.

There was no place to eat anywhere nearby, so we opted for the hotel restaurant. The food was good but the waitress seemed a little odd. I was curious when I heard her accent change drastically each time that she addressed a different customer. It was interesting to watch. She'd have a perfect Aussie accent at one table, then move on and be a southern belle at the next. Either she was having some problems with reality, or she was an actress-to-be, working on different dialects in order to perfect her craft. I'll choose the latter in hopes that she is not offended. It was amusing anyway.

At first the room seemed to be well kept. It appeared clean and neat with everything in order, but the room had a little surprise of its own waiting for us. We had finished dinner, and all of the appropriate phone calls to home. I was getting tired and so was Kim. We each slipped into our respective beds and turned out the lights, then almost simultaneously jumped back out. We were startled and rummaged through the covers and sheets to find out what had pinched us when we laid down. It turned out to be a smattering of field pickers, little brown balls with an outer shell that was consistent with the protective armor of a porcupine. They were strewn throughout both of our beds. Well, I guess it was better than a scorpion or bed bugs, but it made me wonder exactly how well the rooms were cleaned each day.

We finally got everything back in order, with some fresh sheets and a thorough inspection. When Kimberly hit the bed, she passed out in seconds. I took a bit longer, for my thoughts were on the next day's flying and assuring our safe passage through the Rocky Mountains at the Guadalupe Pass. The many pitfalls of flight in mountainous terrain were all that I could think about. It can be treacherous at times. Sudden downdrafts could suck the airplane toward the earth at a rate far greater than it is capable of climbing. Fierce turbulence could bang you around the skies to the point of losing control. You must be wary of roll clouds and high velocity winds aloft. With our type of airplane, instrument flying in the mountains would be even more treacherous and icing would have to be avoided completely, as it would decrease the aircraft's performance so greatly that it could prevent it from being able to fly at all. Hypoxia presents yet another cause for concern. As altitude increases, the density of the air decreases, leaving a pilot less oxygen to breathe. It can impair one's judgment and ability to function. It may even lead to unconsciousness, leaving no one to fly the airplane. Often the symptoms sneak up on you, hidden in feelings of euphoria. The pilot would never know he was a victim until it was too late.

Mountain flight can be beautiful, though, like an aerial view of artwork that was created by the hand of God; yet a pilot must be cautious, especially when flying something as low-powered as our little Cherokee. My apprehensions were justified. I knew the pitfalls and the limitations of our aircraft, but I also knew what to do to minimize the effects they would have on our journey. Early on, when I was planning this flight, I had studied every book and video tape I could find about mountain flying and planned our course to help minimize our exposure to any areas of concern. Nonetheless, we had to cross the Rockies and a smaller mountain range just east of San Diego, or we would never get to the west coast.

My only knowledge of this area of the country came from studying the charts and from the books that I had read by others who had come this way before. One such book made the pass appear as though it were a giant V in the face of torturous cliffs which they had traversed in their Piper Cub on a wing and a prayer. The author portrayed their crossing as a long drawn out battle with upending turbulence that they fought until their muscles fatigued from their attempts to maintain control. His words crossed my mind again and again as I drifted off to sleep, but I had a battle plan of my own.

Chapter 9

Crack! The turbulence hit us again, jarring us up against our seatbelts. Crack! It smashed our little Cherokee in slow rhythmic bursts like a giant hammer from above. We were almost there, though, almost to the pass. Our airplane shuddered and shook so much that the instruments were barely readable. I fought the controls, battling fiercely to keep the sky above and the earth below as mother nature's waves of turbulence tried to roll us inverted. Crack! It hit us again. "Perhaps we should give up, go home," I thought, our little plane can only take so much. If only the winds weren't so fierce. They were blasting through the pass at extremely high velocity as the air had found a path of least resistance. We were caught in that path, the wind was averaging 80 miles per hour on the nose of our airplane. It slowed us to a groundspeed that made a bicycle look fast. Crack! We were struggling to maintain altitude. Then, just as we had entered into the mouth of the pass with no room to turn, we started to descend! I did everything I could to hold on to what altitude we had. I struggled with the controls, hanging the airplane on its propeller just a few knots above the stall speed but to no avail; we continued to sink. We were being sucked into the face of the southern wall of the pass! I remembered my training, "There's always lift at the ridge." I pushed the nose forward in one last ditch attempt to make it to the rising air on the windward side of that ridge. Just as we flew to within twenty feet of the mountain.........

Crack! I woke up from that horrifying dream.

Safe in the confines our little hotel room, I realized that it's not a very good idea to fall asleep to tales of the terrifying flights of other aviators. Anyway, as I said, I had a plan. We would get up early and take the pass in the smooth dense air of the early morning hours. If the winds on either side of the pass were reporting in excess of 20 knots at the surface, we would wait for another day. If the clouds were obscuring the mountains, we would wait as well. We would approach the pass on an angle of 45 degrees and with sufficient altitude to allow us to turn away from the mountains at any given moment and head toward the relative safety of lower terrain. These and other little bits of learning would keep us safe. I was sure of it, despite the terrors of my nightmare. Besides, I had studied the aviation charts and the pass was not truly the wedge in the wall of granite I had pictured. There was considerably more room for maneuvering than that. Any mountain flying should be taken seriously, though, and I would put in place every possible precaution to assure that I kept my sister safe. I had held her life in my hands before, or at least that was the sense of responsibility I had felt during her bone marrow transplant. That time I just sat by and prayed as it was all that I could do. I watched and waited in constant turmoil, hoping that my marrow would do the job. It was a misplaced sense of responsibility.

I understood that on a cognitive level, but it was how I truly felt, responsibility with no authority to effect the outcome. This time it was different. There were variables beyond my control, that is true, but that day when we took to the air to face our challenges, I would have some input, a say in the final result. We were in my arena now, and there would be no room allotted for foolish chances. I was sure that everything would work out right, and our trek across the Rockies would soon be a pleasurable memory for us both.

Kim and I had our coffee in the room, then lugged our belongings downstairs to check out and meet with our ride back to Flower Aviation.

Doug was a manager at the home office of Flower Aviation back in Salinas, Kansas. He was in Hobbs to train Johanna and the rest of the staff at the FBO that his company had recently acquired. My first few words to Doug the day before had nearly sent him into cardiac arrest. He and Kimberly were discussing where we should stay while I was on the phone with the local flight service station. When I came out of the weather room, they had already deemed it best that we stay where Doug had been residing and that he would provide us a ride back to the airport in the morning. Kimberly and Doug happily filled me in on their discussion, and Doug asked what time we would like to start out in the morning. When I requested a five a.m. rendezvous in the hotel lobby, the expression on Doug's face was priceless. He was caught somewhere between, "Are you crazy?" and " If you rea...lly need to." Leaning more heavily toward the, "Are you crazy?" I'm sure. We compromised at six a.m. and that is exactly the time that we met.

Once we had arrived at the FBO, I got the plane ready for departure. Looking across the field, I guessed the visibility to be no more than a mile. Another day, another delay. A few minutes at the weather computer proved my hypothesis and insured our visit to Hobbs would last at least a little while longer. I wanted to be sure that we got underway as soon as possible, so I spent my time checking and rechecking the weather. Meanwhile, Doug, Kimberly and Johanna amused themselves by playing with Mandy, the parrot mascot of Flower Aviation.

Around eight thirty everything was falling into place. The weather had improved all the way to El Paso, and we were packed and ready to go. By nine o'clock we were climbing for altitude, headed for the pass in the calm clear morning air. We were still climbing as we crossed over Carlsbad, New Mexico, and got our first look at the Guadelupe Peak and the pass that shared its name. The land below was nothing but desert, miles and miles of sand pock-marked with sage brush that led us from southeastern New Mexico on into west

Texas.

Just as we passed through 10,000 feet, we grabbed the nasal canulas of our on board oxygen, and I reached into the rear seat for the valve to start the flow. The cockpit was suddenly filled with the overwhelming noise of high pressure oxygen blasting through a broken check valve. It had to be attended to, or our oxygen supply would be depleted long before we began our descent. I gave Kimberly the controls of the airplane. She held our heading in the climb and eventually leveled off at 10,500 feet, our target altitude for the time being. It was only her third time at the controls, and she was handling herself like an old pro. She even responded to a query from air traffic control while I reached over the seat to incur a temporary fix on our oxygen system. Our temporary solution required that we seal off the check valve by hand as we dispersed timed bursts of oxygen every couple of minutes. It wasn't the most efficient solution, but it would do the job and fend off the effects hypoxia. I decided that the smooth air and relatively low surface winds in our vicinity indicated that 10,500 feet would be a sufficient altitude for the pass that morning, and it would reduce our need for supplemental oxygen. After a short while I took over the controls of the aircraft at Kimberly's request, and we started to settle into a routine.

Air temperature is another function of altitude. Most people realize this as they observe the snow covered mountains in the middle of May. For every 1,000 feet of altitude the temperature drops about three and a half degrees Fahrenheit. That little tidbit of science was having an effect on our morning. We were freezing! Level at 10,500 feet in the chilly morning air, we had the heat on and the vent opened just a little, as a precaution. We were calm and relaxed with the pass in our sights, ready for anything, or so we thought. Suddenly, the carbon monoxide monitor started to chirp. I wasn't concerned at first. Dave had said that it would beep from time to time and that it normally ran somewhere around ten parts per million as indicated on the display. I was surprised to see it indicating twenty seven parts per million and climbing! Carbon monoxide poisoning is serious business, and the oxygen tank would only be of minimal assistance if things continued to worsen as the carbon monoxide attaches to the oxygen carrying red blood cells and prevents them from doing their job. All the oxygen in the world won't help if the body can't process it. We shut off the heat. Then we opened every vent in the airplane to try to flush the carbon monoxide from the air. It worked. In seconds the indicator started to drop to a safe operating level. Thank God Dave decided to put a carbon monoxide monitor in his airplane.

We had it checked at the next stop, but the mechanic didn't find anything wrong. We decided to do without the heat and use the vents on every leg just to be sure that everything was safe. I was glad we took that precaution. It wasn't until we had the airplane inspected in California, prior to my return trip with Dave, that they found the minor exhaust leak that was causing the problem.

We had our few little glitches. Nearly every adventure hit's a little snag here and there, but things were going nicely now; everything was falling into place. The weather was cooperating, the airplane was running smoothly with plenty of fuel in the tanks, and we were sailing along at airplane speeds again, just over 105 knots across the ground. All was well in the pass that morning, or I should say over the pass as we were 1,500 feet above its highest ridge.

We were grazing the lower limit of the Rocky Mountains captivated by their beauty. The view alone was worth every ounce of effort we had put into that flight. Kimberly snapped photo after photo of the majestic terrain and majestic is exactly what it was. The peak at Guadalupe rose sharply from the desert floor, piercing the azure sky, its prowess accentuated by the fair weather cumulus clouds that hung in the distance on the edge of the northern horizon. Long steep slopes and sharp cliffs led to the peaks off our right wing.

To our left a lower set of ridges erupted from the windmill laden floor. Beneath us was nothing but desert. I had thought that Kansas and the Oklahoma panhandle were barren, but this area was like the surface of the moon. It was like no terrain that I had ever flown over before. Our gateway to the Southwest was both mysterious and beautiful.

We breezed through the pass that morning as if we were on the wings of angels, in air as smooth as glass, with no headwind and visibility that allowed us to enjoy the wonders that surrounded us. It's true that we had taken every possible precaution, but we were also blessed with weather that was a perfect compliment to our efforts. The pass still held some secrets; Dave and I would experience some of them on our return trip. We faced the challenges of the pass on a different level, crossing it in the evening, but that's a tale for another time.

As for Kimberly and me, we conquered the pass with ease and went on to sail across the playas and deserts of west Texas. We flew over the busy city of El Paso and its international airport, finally descending to 6500 feet after crossing the ridge on the city's western border. We were back to good old pilotage and dead reckoning, following route 10 all the way to the airport at Las Cruces, New Mexico.

Las Cruces was not an originally planned stop. I had chosen it over El Paso the night before while reviewing the charts and realizing the effort it would require for us to climb back up over those ridges to the west of the city. Besides Las Cruces had an interesting name, and I thought it might be a good place to experience. It turned out to be the most enjoyable stop of the entire trip, save our destination.

Approaching the Guadalupe Pass the terrain was stark yet beautiful, almost other worldly.

Crossing the pass went even better than planned on the smooth morning air.

Chapter 10

The airplane we flew was built in Vero Beach, Florida, in the winter of 1968. It was carefully assembled, rivet by rivet, by the dedicated employees of the Piper Aircraft Company. Little did they know that some thirty -five years later, Kimberly and I would be flying it across the country. Nor could they have imagined how well it would still be running.

A few other things were going on that year. Our country was caught up in the turmoil of a war in Vietnam, while on the home front the peace protesters were beginning to get out of hand. It was a difficult year for the United States of America, a year of sadness and political unrest. In the latter part of that year, though, things would start to change. A glimmer of light would shine, as three brave men unified the nation's hopes and dreams, and we all prayed together for their safe return from their journey to the moon.

On December 21, 1968, Colonel Frank Borman, the commander of Apollo 8, and his crew of Jim Lovell and Bill Anders launched on one of the most historic and dangerous missions of the space program. It was a mission of great importance. In the fall of 1968 NASA had received word that our rivals in the space race, the Russians, were planning a lunar fly by before the end of the year.

Deke Slayton had placed a call to Colonel Borman asking him to return to Houston for an emergency meeting. They spoke in private and Colonel Borman was presented with the odds of their mission. The best engineers and scientists in the country could only rate their chance of a successful mission at approximately 33 % . They were given a 33% chance of a failed mission with a safe return to the earth, and a 33% chance that that they would never return. He agreed on the spot to change the itinerary of Apollo 8 from an earth orbit test of the Saturn Five to the first manned flight to the moon. It was a bold decision, but one he made with total confidence, for he felt it was their duty, and he was a man who truly lived the West Point code of Duty, Honor, and Country. He and his team of cold war warriors never even flinched. They just pressed on and did their job.

The first crew to leave the confines of earth orbit did it aboard a rocket booster that had never before been flown on a manned mission. They flew to a destination some two hundred and fifty thousand miles away and tested new theories that held their lives in the balance. They made such a precise reentry into the earth's atmosphere that it was once likened to tossing a letter from a mile away and slipping it through the mail slot in a door. Too steep an angle and they would burn up on reentry, too shallow, and they would skip off into space like a stone skipping across a lake.

I had mentioned earlier my concerns about flying a single engine aircraft in conditions of low ceilings and reduced visibility. These three gentleman would be putting it all on the line, risking everything on the single engine spacecraft that would allow them to break lunar orbit and return to the earth. It was a dangerous mission, but it was the mission that put the Americans in high gear. Like the final kick in a running race, Apollo 8 left the Russians in the dust in the quest for the moon. Its legend lives on in the hearts of those of us who respectfully remember the sacrifices and achievements of America's astronauts. In 1968 I watched and waited with the rest of the country, and I listened as well to the crew of Apollo 8. They read from the book of Genesis with their eyes no doubt fixed on the first view mankind had ever experienced of the earth in its entirety. Little did I know that nearly thirty-five years later I would have the opportunity to meet Colonel Borman on the ramp of Adventure Aviation, in Las Cruces, New Mexico.

When we landed in Las Cruces, our intent was to refuel and continue on to Tucson where we had planned to spend the evening. That never happened. Hurricane Marty had hit the Baja Peninsula and left behind a collection of slow-moving thunderstorms that drifted northwestward across our route of flight. They hung over Deming for two days while we waited things out in Las Cruces. Tucson was not in the cards, at least not on the way to San Diego.

While checking the local flight service station weather report against the information I was receiving from John Millington, I was informed that Colonel Borman lived in Las Cruces. Mr. Millington is the most knowledgeable man I have ever met when it comes to the space race of the 1960's, and he was sure to let me know that I was in the vicinity of one of our nation's greatest aviators.

I just had to ask the lineman if he knew Colonel Borman. After all, I had read his book myself, and I remembered thinking that that he was a true American hero. The lineman told me of Colonel Borman's routine presence on the field and of the airplanes he kept there. He then left with a promise to point out the Colonel if he should happen to drop in while we were still around.

Kimberly and I put 03 Juliet to bed that day with a feeling of accomplishment. We had only completed one leg, from Hobbs to Las Cruces, but it was a significant achievement. We had safely crossed the Rocky Mountains, and I was elated with how well that crossing had gone. I might even have been just a little relieved.

The staff at Adventure Aviation set the tone for our visit to Las Cruces; they were so friendly and helpful. They took an interest in us and our fund raising efforts with a kindness that exceeded our expectations. Kim was speaking with the manager of the FBO after he had queried her about the patch on her shirt. He immediately offered us a fuel discount and a great rate on a rental car, along with some advice as to what was worth seeing in the area. He made us promise to visit White Sands National Park and La Mesilla. While she put together our travel plans, I went off to speak with KimMarie and Mercy, the two women who possessed directions to the best Mexican food in town. While the ladies were drawing me a map, the topic of Colonel Borman came up again. They, too, promised their assistance in the event that an opportunity should arise.

When I went back out to pack up the car, Kimberly was talking to our mother on her cell phone. After mom was informed that we were safely on the ground for the remainder of the day, I called home. I felt bad later, but I cut my call short when Colonel Borman had telephoned the FBO, and the staff informed me that he wished to speak with us. By the time I got inside, he had already decided to come over and say hello. Minutes later, we were shaking hands and discussing our trip. He spoke of a similar adventure that he had enjoyed while ferrying an old Stearman Biplane from coast to coast a few years back. We compared stories for a moment and took a few pictures in front of 03 Juliet; then he offered us the opportunity to visit his hanger and take a peak at the immaculate P-51 Mustang that he flew in air shows. He performs in the Air Force heritage flight, flying formation on an F-15 fighter jet, a duo that combines the military might of modern aviation with the historic prestige of a classic airplane. He didn't say exactly how powerful his P-51 was, but he had mentioned that it was slightly greater than the standard 1300 horsepower. I got a kick out of thinking of this five foot ten, 75 year old man, muscling around that classic warbird in precision flight. I'll be lucky to see 75, let alone accomplish such things.

It was a memorable event, having met Frank Borman. He was one of the great aviation pioneers, a man of courage whose efforts during his career as an astronaut made a significant contribution to the world of flight. I remember reading Colonel Borman's account of his first encounter with Charles Lindbergh; that moment on the ramp in Las Cruces held the same meaning for me. I had read his book and countless others on the space program, and Colonel Borman was a man that I respected, both for his accomplishments and his character. In the few moments I spent with him, it was evident that he was the kind of man I had expected him to be, straight forward and honest, with a commanding presence, even in the latter years of his life. It was an honor to meet him. Looking back, I wish that I had asked him at least a dozen more questions, but I had been caught up in the moment and really didn't want to be a nuisance to him after he had so kindly extended himself for us. He said two things before we parted company that stuck with me. The first was a question, "Which one of you had the cancer?" At that moment I realized how far Kimberly had come. She looked great, the picture of health. Mr. Borman's comment brought that to my attention. The last thing he said to us was, "God Bless." When Kim and I left the airport, my day had been made. It's not often that you get the opportunity to speak with one of your personal heroes.

Meeting Colonel Frank Borman the commander of Apollo 8 and a personal hero of mine.

Getting a look at Colonel Borman"s P-51 Mustang while his mechanic is performing some routine maintenance.

We spent the afternoon sightseeing in the city of Las Cruces, then headed to the White Sands National Park. The road was long and wound its way out of the city toward the craggy peaks in the distance. On a long flat stretch of countryside we witnessed yet another sight that was unique to the Southwest. The dust devil rose like a wispy tornado, spinning some fifty feet into the air and racing across the farmer's field. It spun furiously, then dissipated into thin air before we had the opportunity to take its picture. I wish we could have captured it on film, but I'm glad we had the opportunity to experience it.

It took us just over an hour to get to White Sands and the giant gypsum sand dunes that rose to nearly fifty feet in the air, and it was worth the trip. It was beautiful! Mile after mile of dunes as white as the driven snow were nestled away between the towering mountain peaks while cotton ball clouds rolled across the crisp blue sky of early autumn. It was a visual feast, a view to be savored and reflected upon for years to come. We drank it in slowly as we explored the peaks and valleys of the dunes seeking out the unique flowers and vegetation for the sake of the team photographer. At the top of one of the dunes Kimberly was struck with an idea. She decided to have a little fun. She had observed the tracks in the sand where previous visitors had ridden from the crest of a dune to its base like sleigh riders in a winter wonderland. She carefully picked the tallest dune she could find; it must have been forty or fifty feet high. After positioning me at the base to photograph her moment of grandeur, she climbed to the top. On the count of three, she raised her arms to the heavens and took one giant leap of faith with great hopes of an adventurous slide to the base. Unfortunately, her butt hit the sand and she stopped dead, stuck in the gypsum. As the photographs would later show, she laughed and stumbled her way to the bottom, feeling more amused than victorious. We left White Sands on that comical note and headed straight for the El Comedor restaurant in Old Mesilla.

The Mexican food was excellent. We ate our fill while getting to know the two gentleman sitting next to us. They were a father and son team that had just finished riding their motorcycles back from Colorado. Their description of that long ride through the Rocky Mountains made me add a new item to the things on my personal bucket list.

Kim at the top of a sand dune in White Sands just moments before her misguided leap of faith.

Huge white sand dunes surrounded by majestic mountains make White Sands National Park a feast for the eyes.

A relaxing moment on the square in Old Mesilla.

The photojournalist at work again in a café in Old Mesilla.

Billy the Kid was tried and sentenced in this building but escaped in a dramatic jail break just days later'

It was the first time we had ever heard of the festival of the dead honoring deceased family members but it was a tasteful addition to some of the artisan shops in Old Mesilla.

The next day we returned to Old Mesilla and the collection of craft shops that lined its streets. Old Mesilla can best be described as an artist community, a quaint little town filled with men and women who have mastered the art of pouring a little bit of themselves into everything they do. The artisans fill the local shops with their wares, and the stands in town square give it a festive appeal. It was a nice easy relaxing day spent dining, shopping, and sight seeing. We learned of the historic significance of Old Mesilla which boasted such things as the courthouse where Billy the Kid was sentenced to hang and the oldest building in New Mexico. I bought some gifts for the family and a ring for Kimberly to commemorate our trip. We both pitched in to buy a rosary for our mother from a wonderful woman named Vita.

Vita's husband Max had died in January 2003 of leukemia. She told us of his battle, speaking with great courage, as her eyes filled with tears for the memory of the man she loved. Our conversation with Vita touched our hearts deeply. She and her children have been deprived of a loving husband and father, yet she goes on, bravely facing each new day with thanks for the time they had together. Our prayers go out for her and her family. We asked for a photo of Max, and she gave us one of a special tribute his friends had given him after his death. The soaring hot air balloon with its hanging flag was posted on our website in his memory.

That night I overheard Kimberly crying to her husband Duane on the phone, feeling guilty that she had survived and Max had not. It was more an act of selfless compassion than of true guilt. We will never know why certain people pull through such close brushes with death and others do not. Some would like to place the blame on the victim. They are nothing but religious zealots, claiming that healing had not occurred because of a lack of faith on the part of the stricken individual. What a crock!! We all waver in our faith on a daily basis. Over the past few years I have had my faith torn down. I questioned everything I had ever believed in. When I needed God the most, though, he was always there. Sometimes it was more evident than others, but he was always there.

In the middle of Kimberly's illness, I was so worried about her that I was indeed questioning my faith. You know the drill. "How could this possibly happen to such a good person?" At my darkest hour, though, I received a blessing.

My eldest son Michael has cerebral palsy. During Kim's illness he was assigned to a new van company for his transportation to and from school each day. His driver, Tim Hargett, and I became friends. Tim eventually disclosed to me that he was a Baptist minister who was assisting his father-in-law in his business by driving the daily bus routes that included a stop at our house. Our friendship grew over time, and we went on to many discussions of faith and spirituality which led me through the illness of my sister and many other trials of everyday life. Not everyone is fortunate enough to have a friend who is also a minister appear exactly when he needs him the most, but I was. It is a blessing that I will never forget.

Illnesses such as those faced by Kimberly and Max are devastating. They can either bolster our faith or destroy it. There is, however, one thing that I have learned in the process. Death, as awful as it may seem, is, in fact, a part of life. We will all eventually have to face it. But if we believe in what we profess at church, the human spirit carries on, and our loved ones have another purpose in a life after death. So, it is by the grace of God that some people pass on while others remain here on earth with us. It is his decision, not ours, and it is made with a profound wisdom beyond our human understanding. These thoughts, while consoling, could never fill the void left in the lives of those who have lost the one they love, but time will heal the wounds.

Kimberly and I lost our father twenty years ago. His death is not something that we have forgotten, but over the years our thoughts have changed to memories of the good times we shared rather than the pain of the loss. Every life is precious. Each of us is given a time on this earth that is not much more than a momentary flash in relation to all eternity. But that flash is ours to cherish, to use as we will. I feel that it is far more important that the quality of one's life be measured rather than the quantity. A life well-lived is indeed more treasured than a life of great length but little value. So when we lose the ones we love, it is important to remember what they have brought into our lives, the joys and love that will forever be a part of us. Our fallen loved ones would want nothing less than that we go on with full and rewarding lives, with faith that someday soon the pain of our loss will fade and their presence will remain with us in the fond memories of who it is that they strove with all their might to be.

Chapter 11

Our day had come. The weather from Las Cruces all the way to Yuma, Arizona, was forecast to be CAVU (Clear Above Visibility Unlimited). Our only possible stumbling block was San Diego itself, for it was still reporting very low ceilings and visibilities. It was just under six hours away, though, and we were hoping that it would clear up a bit before our arrival. I decided to put off the decision to proceed to our destination until after we refueled in Casa Grande, Arizona, just north of Tucson.

We departed Las Cruces with mixed emotions. We were glad to be making progress again, but we would both miss that laid back little town filled with hospitality. Las Cruces is a true gemstone, tucked safely away in the mountains of Southern New Mexico. I know someday I will return there for another visit.

On our climb out we caught one last quick look at the giant steel roadrunner that inhabited the rest stop on I-10. We have some photos of that artist's rendition of the desert bird with a "Beware of Snakes" sign in the foreground reminding us to pay attention and remember where we were. Talk about your odd but interesting tourist attractions.

The giant steel road runner made for an odd roadside attraction. The beware of snakes sign reminded us to pay attention to our surroundings.

The interstate was again our primary source of navigation. We flew low at first and then climbed for a better view. Somewhere around Deming we started to feel the effects of the final remnants of Hurricane Marty. This time they were positive effects. We were blazing along at a groundspeed of over 120 knots. Marty had left us a little tailwind, something not often experienced when heading west.

We flew across New Mexico and into Arizona twisting and turning with the interstate. We worked our way across the desert Southwest through the barren valleys with little bursts of life appearing on occasion. I still remember the names, Lourdsburg, Bowie, Wilcox, Benson and the sudden explosion of civilization known as Tucson. In between the cities we didn't see another aircraft flying our route, then suddenly would come the voice of a fellow aviator taking advantage of the pristine Friday morning. Once we passed the city and its little airport, we returned once more to our private tour of the desert.

For years I had wanted to fly in that area of the country. I had been there before, to the city of Phoenix anyway. Each morning I woke up to clear blue skies as far as the eye could see. It was so impressive to me, an aviator from the Northeast where 10 miles visibility is considered good. The Southwest is known for visibility reports of thirty miles, fifty miles, even unlimited. It was so inviting to someone with such a strong desire to fly that he still turns his eyes skyward at the sound of a distant airplane. Unfortunately, on my previous visits, flying was not on the agenda.

Well, Friday, September 26th was to be the day

that my longing was transformed into reality. We were floating along in a river of wind over the buttes and mesas, gazing upon the desert floor, lost in the majesty of the distant mountain peaks. We flew gracefully from valley to valley in the calm smooth air of early morning. It was nothing less than breathtaking, the flight of my dreams. We were both amazed at the beauty of the rugged terrain, and we relished each and every moment of the flight, all the way to Casa Grande, Arizona, our final fuel stop before setting out for San Diego.

I just had to share some of the spectacular desert vistas.

It was a busy day in Casa Grande, and we met some interesting people during our short visit. We spoke with a retired Navy destroyer captain whose skilled craftsmanship had produced a beautiful green and yellow biplane, sure to turn the heads of any passerby. At the soda machine inside, Kimberly made friends with another duo. They were heading east to Roswell, New Mexico, with plans to follow most of our route of flight in reverse. They too flew a Piper Cherokee 140.

We relaxed for a few minutes, then dined on a flight instructor's lunch, peanuts and soda. Anything can be a feast if you're hungry enough. I started making phone calls to check the weather, then spoke with John Millington back in Pennsylvania and with a decision to proceed, I called our friends the Charettes in San Diego.

Just under three hours to go and the journey would be over. It was time to relax, time to revert to flying in the truest sense of the word. No radios, no navigation equipment, just pure barnstorming, low and slow by the seat of your pants. I planned to switch off everything but the transponder, the intercom and the CD player. Configured as such, we could still speak to each other easily and enjoy the musical accompaniment to our visual feast. Now I know the barnstormers never had such amenities, but a chance to listen to Norah Jones while aloft on such a picture perfect day was something I just couldn't resist.

We lifted off from Casa Grande and turned back toward the west. Once clear of the busy airport traffic pattern, I shut everything down and descended to the minimum legal altitude for our flight. If anything went wrong we would simply glide to a landing on the road below, then pick up the cell phone and hope that Verizon could, "hear us now."

We buzzed along like the old timers, able to see and appreciate the wonders of flight. I don't think I ever enjoyed flying as much as I did that day. It was a magic carpet ride, a glorious view of natural wonders set to music like the scenes from a movie.

I eventually decided to work our way back up to 10,500 feet in preparation for picking up our instrument clearance over Yuma. It would be necessary to evoke the support of air traffic control as San Diego was still reporting weather that would require an instrument approach. Southwest of Gila Bend, not yet in the vicinity of Yuma, we were level at our higher altitude but still not talking with ATC. We were still relaxing to the music of Norah Jones. Her "Come Away With Me" CD had become the music of our journey. It will always evoke fond memories. The microphones were pushed off to the side allowing us a break from the interruptions of the sensitive intercom that would both amplify the speaker's voice and cut out the music.

So there we were, flying along to those majestic views, listening to the smooth melodic sounds of my new favorite artist. I really started to relax, even reclined a little, like one might do on a long drive. I was flying by gently nudging the controls with my finger tips while peering over the dashboard to look for traffic. It was so peaceful, so soothing. Then out of nowhere came a sudden shriek from Kimberly. I jumped to attention, looking for another airplane, wondering for a second what the heck she was yelling about. A moment later we were both in an uproar of laughter when we realized the effect of her turning her head into the microphone while belting out the lyrics at the top of her lungs. I don't think we stopped laughing until we were over Yuma.

The weather in Yuma was still clear, and we watched in awe as the expanse of desert passed beneath our wings. It's remarkable how much sand there is out there, endless oceans of it. We were over the interstate, between two huge areas of military airspace, each one a desert used for maintaining the proficiency of our nation's military pilots.

Miles and miles of sand from our perch above Yuma.

Kim at the controls .

We picked up our clearance and descended to 10,000 feet. Everything was going well. We were ready to cross the last mountain range of our journey, just north of the Mexican border, just east of San Diego. We had plenty of altitude, a nice tailwind, more than enough fuel and an airplane that was running like the day she was flown away from the factory. Then ATC decided to issue me a descent to 8,000 feet. Now any pilot with a bit of experience knows that air traffic controllers are an extremely valuable asset, but they are not responsible for the safety of the flight. That responsibility rests with the pilot in command. I wasn't sure at first if I wanted to accept that descent clearance. I was flying a low powered airplane at higher altitudes in very hot and humid air. What that amounts to is a greatly reduced ability to climb. If we started to sink in a downdraft I wanted plenty of room between our airplane and the ridge. I paused for a moment and requested to stay at 10,000 feet. After checking the automated weather reports in the area, I started to feel more comfortable. The winds were light at all reporting stations, and the clouds in the area indicated the air was relatively stable. I decided to take the descent and cross the ridge at 8,000 feet. The turbulence was a little worse, but holding altitude wasn't a real problem. I would later find out that every year or so someone is sucked into that ridge, not paying attention to the limits of his aircraft.

We were almost there, one final obstacle to clear and we were home free, mission complete on terra firma in San Diego. Our last challenge was, as it turned out, much the same as our first challenge back in Pennsylvania during our departure. The weather was again worse than forecasted and refusing to clear. Another coast, another hurricane. Marty had left more moisture in the air than San Diego was capable of dissipating that day.

We crested the ridge and gazed upon a sea of

cloud and fog. The dark gray blanket hung over southern California was one solid sheet of cloud pierced by the high altitude peaks that faded into the coastal plains. I was instructed by ATC to descend once more, and I watched as we sank into the mist, trying to visualize our position relative to the cumulogranite obstacles that surrounded us. I had no choice now but to trust the controllers as they vectored (steered) us around the terrain that we could no longer see. My sight had reverted once more to our trusty panel of instruments.

The airspace was congested, so I decided to shoot a high speed approach, trading altitude for airspeed to keep us from being a nuisance to the twin engine turboprop that was following us in. Everything was coming together perfectly. I was feeling good, comfortable in the environment in which I so often make my living. A pilot from the Northeast spends a good percentage of his time flying in the clouds. I still had my single engine concerns, but our excess of altitude and the slightly higher ceilings put those worries to rest. The controller turned us on to a base leg, then gave us a heading to intercept the final approach course, all the while keeping us in a descent and allowing me the extra airspeed I had wished for. He did a beautiful job providing a nice smooth intercept with the localizer (a device that provides precise horizontal guidance when approaching to land). I eased our descent rate slightly then rode the vertical guidance needles (glide slope) toward the runway. We were racing toward the ground at 140 miles per hour, completely blind in an envelope of cloud, trusting whole heartedly in those invisible radio waves. At 900 feet we were still in the soup with the needles of the instrument landing system nailed in a perfect cross, exactly as they should be. At 700 feet the ground appeared, fading in and out in the wispy bases of the clouds. Then the runway came into view. It was a hazy outline at first that gradually developed into the distinct configuration of our destination airport. We touched down softly and cleared the runway, rolling to a stop. I turned to Kim smiling from ear to ear, and we high fived each other in victory.

We had completed our goal. We had landed safely at the Montgomery Airport in San Diego, California, after some 2400 miles of challenges, 2400 miles that I will forever treasure, for they were the most memorable miles in my more than ten thousand hours of flying. They were the miles that reunited me with my youth and the sister who shared in it. They were the miles that celebrated the miracle of her remission and her return to a full and adventurous life.

Mission Complete on terra firma in San Diego.

Chapter 12

Lt. Colonel Robert J. Charette is a warrior. He has served our country in Afghanistan, Bosnia and both Gulf Wars. He has landed F/A-18 Hornets on the decks of aircraft carriers in the pitch black darkness of moonless nights while his runway, the carrier, pitched and rolled in forty foot swells. He is a man of honor, a United States Marine and a natural born competitor, made for the job of fighter pilot.

Lt. Colonel Charette is now the commanding officer of VMFA 323, The Death Rattlers, stationed in Miramar, California. He is a great man, always true to the Marine Corp Motto of Semper Fidelis. He has always been a faithful and trusted friend. I can attest to that first hand, for we have been friends for over thirty-five years, ever since our days in kindergarten. Bob was always the one kid on the block who wanted to take things a step further, always testing and trying something new. Some things were remarkable successes and others well.......

One time, after returning from the drag races with his father, Bob had taken an interest in the long sleek racing cars known as rails. Their front wheels extended a good distance from the narrow cockpit. The engines were often mounted in the rear and they were fast, really fast. Bob returned with the idea of building a go cart, a downhill racer, along the same lines. We built it together from spare parts that were laying around his father's garage. The chassis was made of wood with wheels from a small bicycle on the rear and those of an old lawnmower on the front. Steering was accomplished by a pivoting board attached to a small frame and two long rails far out in front of our two place cockpit. We used a rope that we found in Bob's back yard for the steering. It was quite a design, or so we thought.

We finished the construction in less than a week and soon found ourselves at the top of the hill on Saginaw Street in a discussion, "You go first!,... No You go first!" We finally came to an agreement and both hopped on for the inaugural run of our new race car. We rolled our prototype into position on the hill that ran between Bob's house and my grandmother's. It was a great hill for go carts. It went straight down for two blocks then turned 90 degrees to the left, allowing us the opportunity to slow to a stop. It was perfect.

We checked for traffic then jumped on, Bob in front and me in the back. In all of our twelve years of life experience, we just couldn't have imagined what would happen next. I kicked off with my left foot, and we both tucked in to help us build speed. We rolled down that hill like a couple of

Olympic bobsled racers. The wheels rolled faster and faster, using gravity to set what had to be a new land speed record. We hit the bottom of that hill in a moment of elation, thrilled with our new go cart and the great velocity we had achieved. Bob pulled hard on the rope that led to our front wheels in an effort to make the corner and roll victoriously to safety. That dry rotted old rope had other ideas. It snapped like a piece of string holding a thousand pound wrecking ball. We sailed past the turn and picked up a little more speed in the last thirty yards of the hill. We went screaming off the cliff at the base of Saginaw Street, flying across an expanse of about twenty feet and landing in the tops of a group of small trees that bent toward the ground with our weight. We came through it all completely unscathed, a true testimony to the saying, “God watches over fools and children.” My guess would be that we qualified for both categories that day.

Many years later Bob would reintroduce me to the world of flight, that time aboard a rented Cessna 152. Sometime after that, when my finances would allow, I started to take flying lessons myself and the rest is history.

It was a regular reunion at the FBO in San Diego. We were greeted by my family, my old friend Bob Charette and his wife Donna and their son John, little Bob was at a school function. We put 03

Juliet away for the weekend and packed our belongings in the rental car. It was great to have completed our mission and even better to be reunited with our family and friends. Kimberly and I returned to the hotel engulfed in a field of questions that told the family of our adventure in the air. That night Bob took us all to dinner at The Chart House, an ocean front restaurant that allowed us to literally fulfill our motto of , "Sea to sea in 2003."

We spent the weekend hanging out with the Charettes, treated to the hospitality for which they are famous. Good friends, family and good food in the sunshine of Southern California; it just doesn't get any better than that.

On Saturday Kimberly took the redeye home, and Dave arrived to fly with me in his trusty Cherokee. I missed Kim when she left. We had shared in something very special and turned another page in our family's history. Our trip brought us closer than we had ever been. It was a special time; it is a treasured memory.

My old friend Bob his wife Donna and their son John

A visit with Bob at work included a chance to fly the F/A-18 Hornet simulator with Chris as my RIO.

Like father like son. (on Bob's deck)

Aunt Kimmie with Chris and Mikey

Kim and I on the Pacific shore outside The Chart-house Restaurant. Sea to sea in 2003!

Epilogue

Wings Across America had started as a dream and ended as a memory of the joys found in the pursuit of that moment of inspiration. It was a celebration of life, family and friends, and most importantly a celebration of the miracle of Kimberly's remission. It was a financial success as well. On December 12, 2003, the Wings Across America team reunited in Philadelphia to present The Leukemia and Lymphoma Society with the check that brought our total to just over 9,000 dollars. That check reflected the generosity of the many people who gave of themselves to help fight blood related cancers, people from all over our great country. If you are reading this, it means that the book got published and that you have helped as well. Fifty percent of every dollar made on this book goes toward the goal of defeating leukemia and helping those afflicted with the deadly disease.

Each night I pray that Kimberly's battle is over for good, that the remission is permanent and her life continues to be the blessed adventure it was meant to be. I also pray a prayer of thanks, for Kimberly's illness taught me a great deal about what is truly important. I look at things a little differently than I used to. I have learned to appreciate the gift of life. It is a gift; we should never forget that. It is up to each of us to cherish that gift and honor it by living well. I have also learned to value the time we are given with the ones we love. They won't be here forever, nor will we. We should make everyday with them count. The true treasures of this world are found in our relationships with those loved ones. And finally, I have learned to appreciate all that I am given. My life has been filled with so many blessings, and I have been granted the eyes with which to see them.

As for our trip, I will always treasure the days that Kimberly and I spent in the skies across America. Those days were nothing less than a gift from God. Three years ago, when it all began, I could never have envisioned such an adventure. That adventure brought us closer to each other and to the country we love. It was a journey that both of us will always remember, the kind of life experience that permeates our existence. Kimberly had asked that I end our last journal entry with the words of Leonardo DaVinci. I'd like to offer them here as well:

"Once you have tasted flight, you will forever walk with your eyes turned skyward, for there you have been and there you will long to return."

God Bless.

Kimberly Furman continues to enjoy a full and rewarding life as a Physician Assistant in York, Pennsylvania. At the time this book was published she is cancer free and enjoying her days and nights with Duane and their daughter Abigail Charlotte.

Michael Rencavage is a pilot for Flexjet, providing leading private jet travel experiences for some of the world's most successful people.

Author's Note

I just wanted to thank everyone who has supported this book, especially Mrs. Barbara Holmes for her hard work in the editing of my original manuscript. I wish to thank each and every person who has donated to Wings Across America and my family and friends for their support. In writing this book I mentioned the friends and co-workers of previous employers; I want to take a moment to honor those individuals who work with me at Flexjet. I am proud to be working among such a fine group of aviation and private travel professionals. I would list some of the great pilots and management personnel and the tremendous people who support our flight crews on a daily basis whom I have had the pleasure to work with, but fear that a failure or oversight on my part would offend someone. That being the case, here is the official shout out to everyone at Flexjet, especially those great aviators with whom I've actually had the chance to share a cockpit.

This book was written during my days at Mid Atlantic Airlines,(Technically USAirways) some ten years ago. Many things have changed since those days but the meaning of the book, its' purpose, remains the same. It is my hope that this book is presented as a thank you to God for all that he has done in my life especially for the blessing of saving my sister Kimberly. My dream is to present The Leukemia and Lymphoma Society with a significant donation each time the royalties from this book are paid to me. I would like to thank you for contributing to the cause by purchasing my book and ask for one additional favor. If you enjoyed the book please recommend it to as many friends as possible. Every purchase provides an additional donation. Besides where else could you fly across the country for just a few dollars and share in the joy and blessing of one woman's recovery from cancer all while donating to a great cause. Thank you again for your support in my fund raising effort. I sincerely hope that you enjoyed the read. If you would like to make a comment or ask a question please feel free to contact me at mikethewriter53@gmail.com

Also, reviews on Amazon where you purchased the book may help sales. Thanks again for your time and support........Mike.

For more information or to make a very appreciated direct donation to The Leukemia and Lymphoma Society please go to their website www.lls.org. It is a great organization whose efforts are truly making a difference in the battle against blood cancers. If you do choose to donate I would request that you put Wings Across America in the box that says in honor of. They have promised to keep track of this and let us know of the impact our efforts have had.

This book is not sponsored or sanctioned by The Leukemia & Lymphoma Society®

Wings Across America is a grass roots fund raising effort that donates fifty cents of every dollar made from the proceeds of this book to The Leukemia and Lymphoma Society. With every Amazon Payday I will write a check for fifty percent of the royalties I receive and send it off to Ms. Caitlin Crowe of The Leukemia and Lymphoma Society as per an agreement I have made with LLS through her. While the book is not sponsored or sanctioned by LLS our fundraising efforts have been approved as a stand alone fundraiser with the proceeds benefitting The Leukemia and Lymphoma Society. Please contact me at mikethewriter53 @gmail.com or caitlin.crowe@lls.org if you have any questions regarding our fundraising efforts.

The photos in this book are best seen in the ebook version on Kindle. If you would like an email containing the photos just send me a request at mikethewriter53@gmail.com

Made in the USA
Middletown, DE
08 October 2022